Lecture Notes in Mathematics

Edited by A. Dold and B. Eckmann

1356

Hans Volkmer

Multiparameter Eigenvalue Problems and Expansion Theorems

Springer-Verlag
Berlin Heidelberg New York London Paris Tokyo

Author

Hans Volkmer
Fachbereich Mathematik, Universität – Gesamthochschule Essen
Universitätsstr. 3, 4300 Essen 1, Federal Republic of Germany

Mathematics Subject Classification (1980): primary: 15A18, 45F99, 47B25, 34B25
secondary: 15A69, 35P10, 46C10

ISBN 3-540-50479-6 Springer-Verlag Berlin Heidelberg New York
ISBN 0-387-50479-6 Springer-Verlag New York Berlin Heidelberg

Printed in Germany

Printing and binding: Druckhaus Beltz, Hemsbach/Bergstr.
2146/3140-543210

PREFACE

Since the publication of Atkinson's book "Multiparameter Eigenvalue Problems, Vol. I" in 1972, multiparameter spectral theory has become a subject of growing interest. Several authors have made important contributions to the theory; see the references at the end of this book. There are also two monographs of Sleeman (1978a) and McGhee and Picard (1988) on this subject.

The contents of the present book can be called "classical multiparameter theory" to distinguish it from more recent developments in multiparameter theory. The book contains the major results of Atkinson's book (most of them are proved differently), several of the contributions made in the last years and some new results. It is thought as a supplement to the excellent book of Atkinson which has the only drawback that we are waiting for volume II for quite a while.

We study two problems: the existence of eigenvalues and the expansion in series of eigenvectors. Each of these problems is treated for multiparameter eigenvalue problems involving (i) Hermitian matrices (ii) compact Hermitian operators, in particular, integral operators (iii) semibounded selfadjoint operators with compact resolvent, in particular, differential operators. Alltogether this leads to six chapters.

It is not assumed that the reader knows already some multiparameter spectral theory but it is supposed that the reader is familiar with the usual one-parameter eigenvalue problems for compact Hermitian operators and their inverses. Brouwer's degree of maps will be used in the first two chapters. Basic properties of the tensor product of linear spaces will be needed in Chapters 4,5,6.

Theorems in multiparameter spectral theory are usually proved under a so-called definiteness condition. It was my aim to prove these theorems under definiteness conditions as weak as possible. For instance, a bounded Hermitian operator A on a Hilbert space H with inner product $\langle\ ,\ \rangle$ is positive definite if $\langle Au,u\rangle$ is positive for all unit vectors u. The operator A is strictly positive definite if there is a positive ε such that $\langle Au,u\rangle \geq \varepsilon$ for all unit vectors u. In such a situation I prefer to assume that A is positive definite even if the proofs would become simpler under strict definiteness. It is this weakening of definiteness conditions which leads to some new results. In particular, the abstract oscillation theorems of the Chapters 2 and 3 under local definiteness in the strong sense and the expansion theorems of the Chapters 5 and 6 under left definiteness are new. More detailed informations on the correspondence of the presented results with those of the literature are given in the Notes at the end of each chapter.

To avoid misunderstandings I have to say that the contents of this small book are far from being complete in any sense. There are many interesting fields in multiparameter theory which are not contained, for example, eigenvalue problems involving non-symmetric operators (see Atkinson (1968) and Isaev (1980)), eigenvalue problems having a continuous spectrum (see Browne (1977a), (1977b), Volkmer (1982) and McGhee and Picard (1988)), nonlinear problems (see Binding (1980b) and Browne and Sleeman (1979b), (1980b), (1981)), indefinite problems (see Binding and Seddighi (1987a) and Faierman and Roach (1987), (1988a)) and .... Perhaps we should add "Vol. I" to the title of the book.

Finally, I wish to thank my colleagues working in multiparameter theory for their stimulation during the last years at various meetings. In particular, I thank Paul Binding and Patrick Browne for several fruitful discussions during my stay at the University of Calgary (March - July 1984). The results of the first three chapters are based on a joint paper with Binding (1986).

Essen, September 1988 Hans Volkmer

CONTENTS

INTRODUCTION

A multiparameter eigenvalue problem contains a finite number of spectral parameters whose number is denoted by k throughout the text. If k is equal to 1 then we obtain an ordinary one-parameter eigenvalue problem. Therefore the simplest "nontrivial" case is when k is equal to 2. The following two-parameter examples illustrate some of the concepts and results of multiparameter spectral theory.

Let us consider the eigenvalue problem

$$\lambda_0 A_{10} x_1 + \lambda_1 A_{11} x_1 + \lambda_2 A_{12} x_1 = 0 \quad , \quad 0 \neq x_1 \in \mathbb{C}^3 \; , \tag{1}$$

$$\lambda_0 A_{20} x_2 + \lambda_1 A_{21} x_2 + \lambda_2 A_{22} x_2 = 0 \quad , \quad 0 \neq x_2 \in \mathbb{C}^2 \; , \tag{2}$$

where

$$A_{10} = \begin{pmatrix} 4 & 0 & 0 \\ 0 & 0 & 0 \\ 0 & 0 & 0 \end{pmatrix} , \quad A_{11} = \begin{pmatrix} 1 & 0 & 0 \\ 0 & 6 & 0 \\ 0 & 0 & 1 \end{pmatrix} , \quad A_{12} = \begin{pmatrix} 0 & 1 & 0 \\ 1 & 0 & 1 \\ 0 & 1 & 0 \end{pmatrix} ,$$

$$A_{20} = \begin{pmatrix} 20 & 0 \\ 0 & 0 \end{pmatrix} , \quad A_{21} = \begin{pmatrix} 0 & \sqrt{3} \\ \sqrt{3} & 0 \end{pmatrix} , \quad A_{22} = \begin{pmatrix} 7 & 0 \\ 0 & 1 \end{pmatrix} .$$

The eigenvalues are nonzero tuples $\lambda = (\lambda_0, \lambda_1, \lambda_2)$ of complex numbers such that (1) and (2) can be solved simultaneously. It should be observed that an "eigenvalue" is a vector and not a scalar. The notion "eigenvector" is reserved for the tensor products of the solutions x_1 and x_2 of (1) and (2), respectively. There are three spectral parameters $\lambda_0, \lambda_1, \lambda_2$ but they only count for two because (1),(2) is written in a homogeneous formulation i.e. if λ is an eigenvalue then also $\alpha\lambda$ is one for each nonzero complex number α. Our example is so simple that we can calculate the eigenvalues explicitely. Of course, this will be possible only in a very limited number of examples. We first determine the values of $\lambda_0, \lambda_1, \lambda_2$ such that the matrix

$$\lambda_0 A_{10} + \lambda_1 A_{11} + \lambda_2 A_{12} \tag{3}$$

becomes singular. These values are given by

$$(4\lambda_0 + 2\lambda_1)\,\lambda_2^2 = 6(4\lambda_0 + \lambda_1)\,\lambda_1^2 \; . \tag{4}$$

Similarly, the matrix

$$\lambda_0 A_{20} + \lambda_1 A_{21} + \lambda_2 A_{22} \tag{5}$$

is singular if

$$3\lambda_1^2 = 20\,\lambda_0\,\lambda_2 + 7\lambda_2^2 \; . \tag{6}$$

The eigenvalues $\lambda = (\lambda_0,\lambda_1,\lambda_2)$ are the simultaneous solutions of (4) and (6). It is easily seen that λ_0 is nonzero for eigenvalues $(\lambda_0,\lambda_1,\lambda_2)$. Hence we can go over from the homogeneous formulation of the eigenvalue problem (1),(2) to an inhomogeneous one by setting $\lambda_0 = 1$. Now the eigenvalues $(1,\lambda_1,\lambda_2)$ have real components which can be determined by a picture.

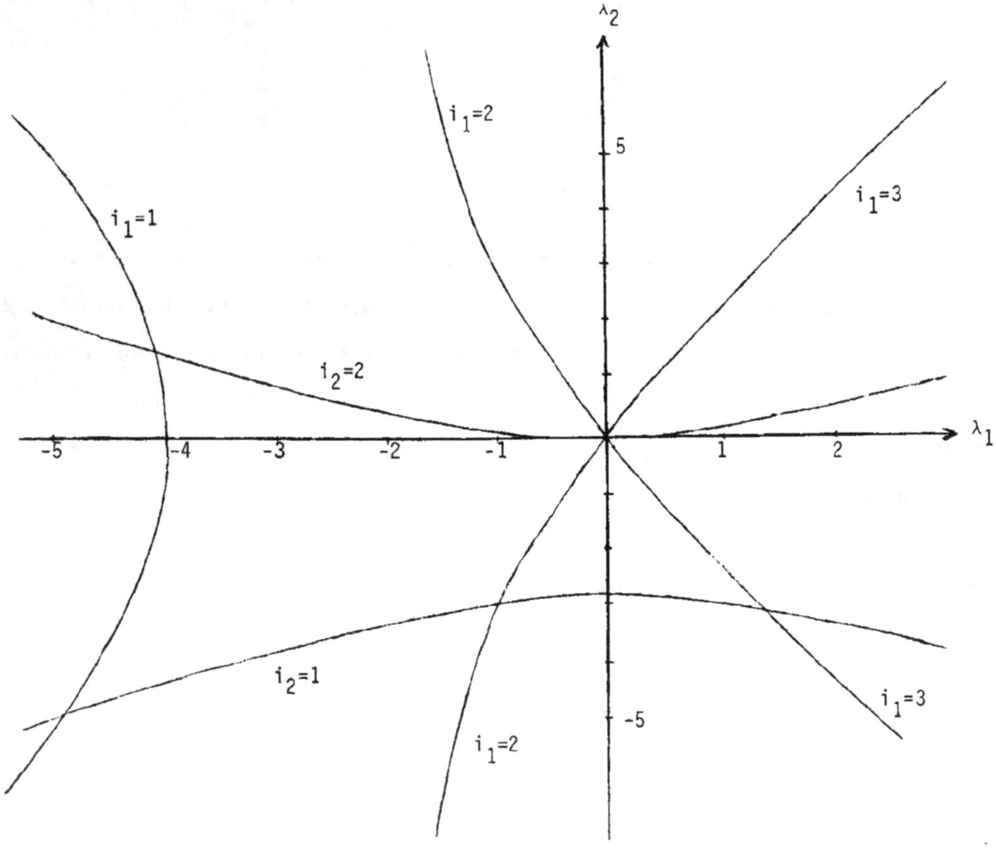

The curves indicated by $i_1 = 1$, $i_1 = 2$, $i_1 = 3$ consist of those pairs (λ_1,λ_2) for which the i_1th greatest eigenvalue of (3), counted according to multiplicity, is equal to 0 . Similarly, the curves indicated by $i_2 = 1$, $i_2 = 2$ consist of the pairs (λ_1,λ_2) for which the i_2th greatest eigenvalue of (5) is 0 . The eigenvalues of problem (1),(2) correspond to the points of intersection of these eigencurves. For example, (-5,-5) lies on the curves $i_1 = 1$ and $i_2 = 1$. We say that the eigenvalue (1,-5,-5) has index $(i_1,i_2) = (1,1)$. The eigenvalue (1,0,0) has two indices, namely $(i_1,i_2) = (2,2)$ and $(i_1,i_2) = (3,2)$. We say that the eigenvalue (1,0,0) has multiplicity 2 .

We see that, for every given index (i_1,i_2) , $i_1 = 1,2,3$, $i_2 = 1,2$, there is exactly one eigenvalue $(1,\lambda_1,\lambda_2)$ which has index (i_1,i_2) . This is a special case of the statement of Theorem 1.4.1 because our eigenvalue problem (1),(2) is definite with respect to (1,0,0) i.e. we have

$$\det \begin{pmatrix} 1 & 0 & 0 \\ \langle A_{1o}u_1,u_1\rangle_1 & \langle A_{11}u_1,u_1\rangle_1 & \langle A_{12}u_1,u_1\rangle_1 \\ \langle A_{2o}u_2,u_2\rangle_2 & \langle A_{21}u_2,u_2\rangle_2 & \langle A_{22}u_2,u_2\rangle_2 \end{pmatrix} > 0 ,$$

for all unit vectors $u_1 \in H_1$ and $u_2 \in H_2$, where $\langle\ ,\ \rangle_1$ and $\langle\ ,\ \rangle_2$ are the usual inner products in $\mathbb{C}^3$ and $\mathbb{C}^2$, respectively.

There is also an expansion theorem for the eigenvalue problem (1),(2). For every eigenvalue $(1,\lambda_1,\lambda_2)$, we choose solutions x_1,x_2 of (1),(2), respectively. For the eigenvalue (1,0,0) of multiplicity 2 , we take two linear independent solutions x_1 of (1). Then we obtain 6 decomposable tensors $x_1 \otimes x_2$ which form a basis of the tensor product $\mathbb{C}^3 \otimes \mathbb{C}^2$. This is a special case of the results of Section 4.5.

Multiparameter eigenvalue problems arise in mathematical physics if the method of separation of variables can be used to solve boundary eigenvalue problems. Usually, we have a partial differential equation in k independent variables which contains one spectral parameter. If it is possible to apply the method of separation of variables to solve the equation then we obtain k ordinary differential equations linked by k-1 separation constants and the original spectral parameter. Together

Section 3 is devoted to the study of diagonal forms. Most of its results may be found (explicitly or implicitly) in [L] or in [C]; we however give an essentially self-contained presentation of the results and proofs.

For the diagonal forms, finite-dimensional generalized Clifford algebras may be constructed. Just as for quadratic forms one obtains these Clifford algebras as tensor products of cyclic algebras. More generally, suppose f_1 and f_2 are forms of degree d in disjoint sets of variables, and let C_i be a generalized Clifford algebra for f_i $(i = 1, 2)$. We assume that k contains a d-th primitive root ξ of unity. Then we define the $\mathbf{Z}/d\mathbf{Z}$-graded tensor product $C_1 \widehat{\otimes} C_2$ as the ordinary tensor product equipped with the multiplication defined by $(a \otimes b)(c \otimes d) = \xi^{(\deg b)(\deg c)} ac \otimes bd$ for homogeneous elements $b \in C_2$ and $c \in C_1$. It turns out (theorem 3.1) that $C_1 \widehat{\otimes} C_2$ is a generalized Clifford algebra for $f_1 + f_2$. Now if $f = a_1 X_1^d + \ldots a_n X_n^d$ is a diagonal form with $a_i \in k$, $a_i \neq 0$ for $i = 1, \ldots, n$, then $C_i = k[e_i]/(e_i^d - a_i)$ is a generalized Clifford algebra for $a_i X_i^d$, whence $C(f, \xi) = C_1 \widehat{\otimes} \ldots \widehat{\otimes} C_n$ is a generalized Clifford algebra for f, whose dimension over k is d^n. The structure of this algebra can be described quite easily. In theorem 3.6 it is shown that $C_0(f, \xi)$ is simple if n is odd and semisimple if n is even. The consequences for matrix factorizations of diagonal forms are formulated in theorem 3.9. At the end of this section we work out explicit factorizations of $\sum_{i=1}^{n} X_i^d$ over $\mathbf{C}$.

Finally, in section 4 we show that a linear matrix factorization $f = \alpha_0 \cdot \ldots \cdot \alpha_{d-1}$ corresponds to a free module F over the hypersurface ring $R = S/(f)$ together with a filtration of F, whose quotients are linear MCM-modules over R. In particular, together with the results of section 3, it follows that a hypersurface ring of a diagonal form admits linear MCM-modules.

Many questions remain open [but see the STOP PRESS!]. We list a few of them:

1) Does every (homogeneous) form admit a finite-dimensional generalized Clifford algebra?
2) Do the linear MCM-modules together with R generate the Grothendieck group of R?
3) Can the periodicity theorem of Knörrer [K] be generalized to forms of higher degree?
4) Which forms can be transformed into diagonal forms?

We wish to thank T. G. Ivanova with whom we had many stimulating and helpful discussions, and P. M. Cohn for his valuable comments and suggestions. We also thank Bokut who informed us that L'vov and Nesterenko (answering a question of Krendelev) reported on the solution of question 1 at the 17:th All Union Algebra Conference in Minsk 1983, and announced this and related results (without proofs) in the Proceedings of that conference (pp 118 and 137, in Russian).

In particular, we thank the referee for putting our attention to the extensive work already done concerning generalized Clifford algebras (e.g. in [C], [H], [L], and [R]).

Finally we would like to express our gratitude to the organizers of the Fifth National School in Algebra in Varna, who brought together two of the authors of this paper and made possible many fruitful discussions with other participants of this conference that were indispensable for writing this paper.

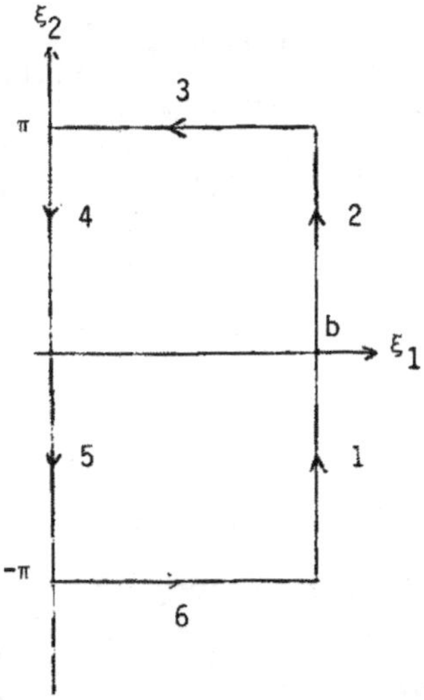

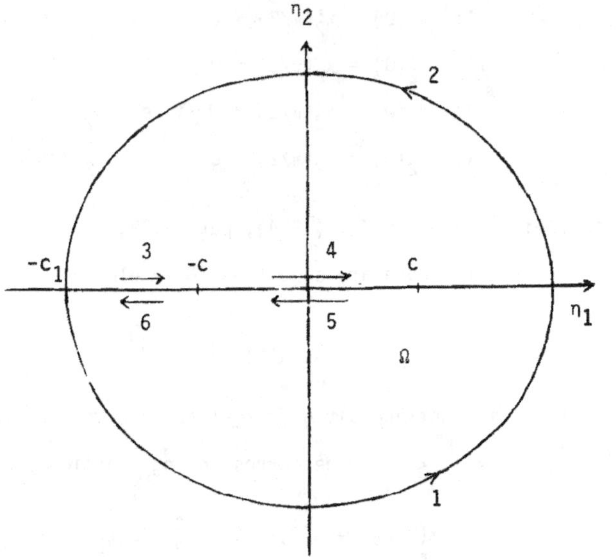

Setting

$$y(\eta_1,\eta_2) = x(\xi_1,\xi_2) \quad ,$$

the equation (7) transforms into

$$\frac{\partial^2 x}{\partial \xi_1^2} + \frac{\partial^2 x}{\partial \xi_2^2} + \frac{c^2 \nu}{2} (\cosh 2\xi_1 - \cos 2\xi_2)\, x = 0 \quad .$$

If we assume a solution of this equation of the form

$$x(\xi_1,\xi_2) = x_1(\xi_1)\, x_2(\xi_2) \quad ,$$

we obtain for x_1 and x_2 the ordinary differential equations

$$x_1'' + (2\lambda_1 \cosh 2\xi_1 - \lambda_2)\, x_1 = 0 \quad , \tag{9}$$

$$x_2'' - (2\lambda_1 \cos 2\xi_2 - \lambda_2)\, x_2 = 0 \quad , \tag{10}$$

where λ_2 is the separation constant and $4\lambda_1 = c^2\nu$. Equation (10) is Mathieu's differential equation and (9) is its modified form. The function x_2 has to be a periodic function with period 2π . The solutions of (10) with period 2π are usually divided into the following four sets.

$$\left.\begin{array}{llll} \text{I:} & x_2'(0) = x_2'(\pi/2) = 0 & \text{i. e.} & x_2 \text{ is even and of period } \pi\ , \\ \text{II:} & x_2'(0) = x_2(\pi/2) = 0 & \text{i. e.} & x_2 \text{ is even and of half-period } \pi\ , \\ \text{III:} & x_2(0) = x_2'(\pi/2) = 0 & \text{i. e.} & x_2 \text{ is odd and of half-period } \pi\ , \\ \text{IV:} & x_2(0) = x_2(\pi/2) = 0 & \text{i. e.} & x_2 \text{ is odd and of period } \pi\ ; \end{array}\right\} \quad (11)$$

see [Meixner and Schäfke (1954), pages 108, 149].

Further, the boundary condition (8) yields

$$x_1(b) = 0 \ . \qquad (12)$$

Since y is continuously differentiable in the neighborhood of the focal line $(\eta_1,0)$, $-c \leq \eta_1 \leq c$, the corresponding function x satisfies

$$x(0,\xi_2) = x(0,-\xi_2)\ , \quad \frac{\partial x}{\partial \xi_1}(0,\xi_2) = -\frac{\partial x}{\partial \xi_1}(0,-\xi_2) \ .$$

Hence we obtain the boundary conditions

$$\left.\begin{array}{lll} x_1'(0) = 0 & \text{in the cases} & \text{I , II ,} \\ x_1(0) = 0 & \text{in the cases} & \text{III , IV .} \end{array}\right\} \qquad (13)$$

In each of the four cases I, II, III, IV we have thus obtained a two-parameter Sturm-Liouville eigenvalue problem (9) , (10), (11), (12), (13). For example, in case I our eigenvalue problem consists in finding values of λ_1 and λ_2 such that equation (9) has a nontrivial solution satisfying the boundary conditions $x_1'(0) = 0$, $x_1(b) = 0$ and, simultaneously, (10) has a nontrivial solution satisfying $x_2'(0) = 0$, $x_2'(\pi/2) = 0$. If (λ_1,λ_2) is an eigenvalue then $\nu = 4\lambda_1/c^2$ is an eigenvalue of the original eigenvalue problem (7), (8). Using expansion theorems for the two-parameter eigenvalue problems, we can also show that the converse statement is true: if ν is an eigenvalue of (7), (8) then there exists an eigenvalue (λ_1,λ_2) of one of the four two-parameter problems such that $\nu = 4\lambda_1/c^2$.

There holds the following result concerning the existence of eigenvalues in each of the four cases. For each pair n_1 , n_2 of nonnegative integers, there are uniquely determined real numbers λ_1 , λ_2 such that (9) admits a solution satisfying the boundary conditions (12), (13) and having exactly n_1 zeros in the open interval $]0,b[$ and such that (10) admits a solution satisfying the boundary conditions (11)

THEOREM 1.3. *Assume that k is infinite. Let $f \neq 0$ be a homogeneous polynomial of degree d.*

i. *The equivalence classes of matrix factorizations of f correspond bijectively to the isomorphism classes of $\mathbf{Z}/d\mathbf{Z}$-graded modules over the universal Clifford algebra of f.*

ii. *Let $M = \bigoplus_{i=0}^{d-1} M_i$ correspond to the matrix factorization $f = \alpha_0 \cdot \ldots \cdot \alpha_{d-1}$. Then*

1) $\dim_k M_i$ is equal to the size of the matrices α_i for all i.

2) If $j \in \mathbf{Z}$, then the shifted module $M(j)$ corresponds to the matrix factorization $f = \alpha_j \cdot \alpha_{j+1} \cdot \ldots \cdot \alpha_{j+d-1}$.

3) This matrix factorization is decomposable if and only if M is decomposable.

Proof. We just indicate how a $\mathbf{Z}/d\mathbf{Z}$-graded module $M = \bigoplus_{i=0}^{d-1} M_i$ defines a matrix factorization of f. Choose a $u \in V$ such that $f(u) \neq 0$; u is a unit in $C(f)$. Since $u \in C_1(f)$, the multiplication by u induces k-isomorphisms u: $M_i \xrightarrow{\sim} M_{i+1}$ for $i = 0, \cdots, d-1$, whence all M_i have the same k-vectorspace dimension. This implies that the k-linear maps ϕ_i: $V \to \mathrm{Hom}_k(M_i, M_{i+1})$ for $i = 0, \cdots, d-1$ define square matrices α_i of linear forms (with respect to some bases of the M_i). Clearly $f = \alpha_0 \cdot \ldots \cdot \alpha_{d-1}$. □

We now describe the algebra $C(f)$ more precisely: For any $\ell \geq 0$, let N_ℓ be the set of n-tuples $\nu = (\nu_1, \ldots, \nu_n)$ with $\nu_i \geq 0$ for $i = 1, \ldots, n$ and $\sum_{i=1}^n \nu_i = \ell$. Let $N = \bigcup_{\ell \geq 0} N_\ell$. As usual, we set $x^\nu := x_1^{\nu_1} \cdot \ldots \cdot x_n^{\nu_n}$. Then (for some $a_\nu \in k$) we have

$$f = \sum_{\nu \in N_d} a_\nu x^\nu \; .$$

Let $\nu \in N$; a monomial in the generators $e_1, \ldots, e_n$ is said to have *multidegree* ν, if e_i occurs exactly ν_i times as a factor in this monomial for $i = 1, \ldots, n$. For example, the monomials of multidegree $(2,1)$ are $e_1^2 e_2$, $e_1 e_2 e_1$, and $e_2 e_1^2$.

We let g_ν be the sum of all monomials of multidegree ν, so that for instance $g_{(2,1)} = e_1^2 e_2 + e_1 e_2 e_1 + e_2 e_1^2$. For convenience we put $g_\nu = 0$ if ν is an n-tuple *not in* N, so that for instance $g_{(4,-1)} = 0$.

Let $J(f)$ be the two-sided ideal of T generated by the elements $g_\nu - a_\nu$, $\nu \in N_d$, and let $S(f) = T/J(f)$. Then we have

LEMMA 1.4.

i. $I(f) \subseteq J(f)$

ii. $I(f) = J(f)$, *if k is infinite.*

Proof. If $x = \sum_{i=1}^n x_i e_i \in V$, then

$$x^{\otimes d} - f(x_1, \ldots, x_n) = \sum_{\nu \in N_d} (g_\nu - a_\nu) x^\nu \; .$$ □

In other words, there is a natural epimorphism $C(f) \twoheadrightarrow S(f)$, which is an isomorphism if k is infinite.

Next we shall employ the "Diamond lemma" techniques (cf [Be]), in order to study the ideal $J(f)$.

If we set $e_1 < e_2 < \ldots < e_n$, we can order the monomials of T in the e_i in the standard way: first by length, then (for monomials of the same length) lexicographically.

CHAPTER 1

MULTIPARAMETER EIGENVALUE PROBLEMS FOR HERMITIAN MATRICES

1.1 Introduction

We suppose given k linear spaces H_r, $r = 1,\dots,k$, all over the complex field, nonzero, and of finite dimension. In each H_r there is an inner product $\langle\ ,\ \rangle_r$ with associated unit sphere U_r. For each r, let A_{rs}, $s = 0,\dots,k$, be a set of $k+1$ Hermitian operators on H_r.

We shall study the multiparameter eigenvalue problem

$$\sum_{s=0}^{k} \lambda_s A_{rs} u_r = 0, \quad u_r \in U_r, \ r = 1,\dots,k . \tag{1.1.1}$$

We shall use the term *eigenvalue* to denote a nonzero $(k+1)$-tuple of scalars $\lambda = (\lambda_0,\dots,\lambda_k)$ such that there exist vectors $u_r \in U_r$, $r = 1,\dots,k$, satisfying the k equations (1.1.1).

In most cases the eigenvalue problem (1.1.1) will be treated under the hypothesis of local definiteness. We call (1.1.1) *locally definite* if the real k by $k+1$ matrix

$$W(u) := \begin{pmatrix} \langle A_{10}u_1,u_1\rangle_1 & \cdots & \langle A_{1k}u_1,u_1\rangle_1 \\ \vdots & & \vdots \\ \langle A_{k0}u_k,u_k\rangle_k & \cdots & \langle A_{kk}u_k,u_k\rangle_k \end{pmatrix} \tag{1.1.2}$$

is of maximal rank for all $u = (u_1,\dots,u_k) \in U := U_1 \times \dots \times U_k$, i. e.

$$\operatorname{rank} W(u) = k \quad \text{for all} \quad u \in U . \tag{1.1.3}$$

In the next section we show that the eigenvalues of a locally definite problem (1.1.1) can be indexed in a natural manner, and in Theorem 1.2.3 we prove that the eigenvalues are uniquely determined by a signed index. In Section 1.3 we provide the basic properties of Brouwer's degree of maps which we need in order to prove Theorem 1.4.1 on the existence of eigenvalues of given signed index. Subsequently, we note two corollaries of the existence theorem, a perturbation theorem and a criterion for the problem (1.1.1) to be locally definite. In Section 1.5 we consider definite problems. In Section 1.6 we refer to the literature.

1.2 Indexed eigenvalues

We consider the eigenvalue problem (1.1.1). For given $\lambda = (\lambda_0,\ldots,\lambda_k) \in \mathbb{R}^{k+1}$ and $r = 1,\ldots,k$, we list the eigenvalues of the Hermitian operator

$$\sum_{s=0}^{k} \lambda_s A_{rs} \tag{1.2.1}$$

in decreasing order, according to multiplicity, as

$$\rho_r(\lambda,1) \geq \rho_r(\lambda,2) \geq \ldots \geq \rho_r(\lambda, \dim H_r) \quad . \tag{1.2.2}$$

Then a nonzero $\lambda \in \mathbb{R}^{k+1}$ is an eigenvalue of (1.1.1) if and only if there exists a multiindex

$$i = (i_1,\ldots,i_k) \quad , \quad i_r = 1,\ldots,\dim H_r \ , \quad r = 1,\ldots,k \ , \tag{1.2.3}$$

such that

$$\rho_r(\lambda,i_r) = 0 \quad \text{for every} \quad r = 1,\ldots,k \quad . \tag{1.2.4}$$

If the equations (1.2.4) are satisfied then we say that the eigenvalue λ has index i. In general, an eigenvalue can have several indices. The number of these indices is called the *multiplicity* of the eigenvalue.

The maximum-minimum-principle for the eigenvalues of the Hermitian operators (1.2.1) yields the useful representation

$$\rho_r(\lambda,i_r) = \max\{\min\{w_r(u_r)\lambda \mid u_r \in F_r \cap U_r\} \mid F_r \subset H_r, \dim F_r = i_r\} \ , \tag{1.2.5}$$

where $w_r(u_r)$ denotes the rth row of the matrix (1.1.2) and $w_r(u_r)\lambda$ is the usual product of a row and column vector. Here and in the sequel we consider λ as a column vector. For the maximum-minimum-principle we refer to [Weinstein and Stenger (1972), Chapter 2]. There it is called minimum-maximum-principle because the eigenvalues are listed in increasing order.

We now turn to consequences of local definiteness of the eigenvalue problem (1.1.1). Let us first express the rank condition (1.1.3) by determinants as usual. For $u \in U$ and $s = 0,\ldots,k$, we denote by $(-1)^s \delta_s(u)$ the determinant of the matrix $W(u)$ with s^{th} column deleted. Then the eigenvalue problem is locally definite if and only if the column vector

$$\delta(u) := (\delta_0(u),\ldots,\delta_k(u)) \text{ is nonzero for all } u \in U \ . \tag{1.2.6}$$

We also know from the elementary theory of matrices that local definiteness is equivalent to the condition that

$$\text{Ker } W(u) = \{\alpha\ \delta(u) \mid \alpha \text{ complex}\} \quad \text{for all} \quad u \in U \ , \tag{1.2.7}$$

where Ker $W(u)$ denotes the kernel of the matrix $W(u)$. In the rest of this section we shall assume that (1.1.1) is locally definite.

If λ is an eigenvalue then the equations (1.1.1) hold for some $u = (u_1,\ldots,u_k)$. It follows immediately that $W(u)\lambda = 0$. Hence, by (1.2.7), λ is a complex multiple of $\delta(u)$, a vector which has real components. Therefore, and since the eigenvalue problem (1.1.1) is homogeneous in λ , it will be sufficient to search eigenvalues in the unit sphere S^k of $\mathbb{R}^{k+1}$.

We now define three subsets of S^k which will play the crucial role in this chapter:

$$P \ := \{\lambda \in S^k \mid W(u)\lambda = 0 \ \text{ for some } \ u \in U\} \ ,$$

$$P^+ := \{\delta(u) \ / \ \|\delta(u)\| \mid u \in U\} \ ,$$

$$P^- := \{-\delta(u) \ / \ \|\delta(u)\| \mid u \in U\} = -P^+ \ ,$$

where $\| \ \|$ denotes the Euclidean norm in $\mathbb{R}^{k+1}$. We note that P contains all eigenvalues of (1.1.1) lying in S^k .

LEMMA 1.2.1. *Assume that the eigenvalue problem (1.1.1) is locally definite. Then* P *is the disjoint union of the compact and arcwise connected sets* P^+ *and* P^- .

Proof. It follows from (1.2.7) that P is the union of P^+ and P^- . In order to prove that P^+ and P^- are disjoint, let $\lambda \in P$ and let Y be the set of all $u \in U$ such that $W(u)\lambda = 0$. This set is a product $Y = Y_1 \times \ldots \times Y_k$, where

$$Y_r = \{u_r \in U_r \mid w_r(u_r)\lambda = 0\} \ .$$

The following general Lemma 1.2.2 shows that Y_r is arcwise connected for every r. Hence Y is arcwise connected, too. Now there is a continuous real-valued function α on Y such that $\lambda = \alpha(u)\delta(u)$ for all $u \in Y$. Since Y is connected and α has no zeros, α has constant sign on Y. This proves that λ cannot belong to both P^+ and P^-.

The sets P^+ and P^- are continuous images of the compact and arcwise connected set U. Hence P^+ and P^- are compact and arcwise connected, too. We note that U is compact because the spaces H_r are finite dimensional, and that U is arcwise connected because of Lemma 1.2.2.□

LEMMA 1.2.2. *Let* H *be a complex linear space with inner product* $\langle\ ,\ \rangle$ *and unit sphere* U. *Let* ψ *be a Hermitian sesquilinear form on* H. *Then the set*

$$Y = \{u \in U \mid \psi(u,u) = 0\}$$

is arcwise connected. In particular, U *is arcwise connected.*

Proof. Let $x,y \in Y$. If x,y are linearly dependent then there is a real number θ such that $y = \exp(i\theta)x$. Then the continuous path $\exp(it\theta)x$, $0 \le t \le 1$, connects x and y within Y. Now let x,y be linearly independent. Then we choose the real number θ such that the real part of $\exp(i\theta)\psi(x,y)$ vanishes. The segment

$$z(t) = t\exp(i\theta)x + (1-t)y, \quad 0 \le t \le 1 ,$$

connects y and $\exp(i\theta)x$, does not cross 0 and satisfies $\psi(z(t), z(t)) = 0$ for all $0 \le t \le 1$. Hence the continuous path $z(t) / \langle z(t), z(t)\rangle^{1/2}$ connects y and $\exp(i\theta)x$ within Y. By what we have shown in the first part of the proof, we can also connect x and y by a continuous path within Y. This shows that Y is arcwise connected. Choosing $\psi = 0$, we see that U is arcwise connected. □

If λ is in P^+ or P^- then we say that λ has *signum* $+1$ or -1, respectively. We now are in a position to prove the uniqueness of an eigenvalue of given signed index (i,σ), i. e. an eigenvalue which has index i and signum σ.

THEOREM 1.2.3. *Let the eigenvalue problem (1.1.1) be locally definite. Then there is at most one eigenvalue in* S^k *of given signed index.*

Proof. We suppose that λ and μ are two distinct eigenvalues in S^k of the same index i. We shall show that λ and μ have different signum. By (1.2.4) and (1.2.5), there are subspaces F_r of H_r of dimension i_r such that

$$0 = \rho_r(\lambda, i_r) = \min\{w_r(u_r)\lambda \mid u_r \in F_r \cap U_r\} \quad . \tag{1.2.8}$$

Now if u_r' minimizes $w_r(u_r)\mu$ over $u_r \in F_r \cap U_r$ then (1.2.8) and (1.2.4), (1.2.5) with μ in place of λ yield

$$W(u')\lambda \geq 0 \geq W(u')\mu \quad , \tag{1.2.9}$$

where we interpret vector inequalities componentwise. Similarly, there is $u'' \in U$ such that

$$W(u'')\mu \geq 0 \geq W(u'')\lambda \quad . \tag{1.2.10}$$

Since λ and μ are distinct,

$$\pi(t) := (t\lambda-(1-t)\mu) / \| t\lambda-(1-t)\mu\| \, , \quad 0 \leq t \leq 1 \quad ,$$

defines a continuous path in S^k joining $-\mu$ to λ. From (1.2.9) and (1.2.10) it follows that

$$W(u')\pi(t) \geq 0 \geq W(u'')\pi(t) \quad \text{for all} \quad 0 \leq t \leq 1 \quad .$$

Obviously, the set $\{w_r(u_r)\pi(t) \mid u_r \in U_r\}$ is an interval for every r and t. Hence there are vectors $u(t) \in U$ such that

$$W(u(t))\pi(t) = 0 \quad \text{for all} \quad 0 \leq t \leq 1 \quad .$$

This shows that the path π is contained in P. By Lemma 1.2.1, the path π is contained entirely in P^+ or P^-. Therefore $\pi(0) = -\mu$ and $\pi(1) = \lambda$ have the same signum and, consequently, μ and λ have different signum.□

1.3 The degree of a map

Let Ω be a bounded open subset of $\mathbb{R}^k$, and let $\overline{\Omega}$ denote its closure. Let $f : \overline{\Omega} \to \mathbb{R}^k$ be a continuous map which is continuously differentiable on Ω.

Further let y be a point in $\mathbb{R}^k$ such that the equation $f(x) = y$ has no solutions x lying in the boundary $\partial\Omega$ of Ω. Then, if the Jacobian determinant $\det f'(x)$ is nonzero for all $x \in \Omega$ satisfying $f(x) = y$, the degree $\deg(f,\Omega,y)$ is defined by

$$\deg(f,\Omega,y) := \sum_{f(x)=y} \operatorname{sign} \det f'(x) , \qquad (1.3.1)$$

where the sum is taken over the finite number of $x \in \Omega$ solving $f(x) = y$. The definition of the integer $\deg(f,\Omega,y)$ can be extended to every continuous map $f : \overline{\Omega} \to \mathbb{R}^k$ and every $y \in \mathbb{R}^k$ satisfying $y \neq f(x)$ on $\partial\Omega$. We need not know the details of this more general definition. We only need the following two basic properties of the degree.

THEOREM 1.3.1. *(i) Let* $f : \overline{\Omega} \to \mathbb{R}^k$ *be a continuous map, and let* y *be different from* $f(x)$ *for all* $x \in \partial\Omega$. *Then, if* $\deg(f,\Omega,y)$ *is nonzero, there exists at least one solution* $x \in \Omega$ *of the equation* $f(x) = y$.
(ii) Let $g : [0,1] \times \overline{\Omega} \to \mathbb{R}^k$ *be continuous and assume that* y *is different from* $g(t,x)$ *for all* $(t,x) \in [0,1] \times \partial\Omega$. *Then* $\deg(g(t,.),\Omega,y)$ *is independent of* t *for* $t \in [0,1]$.

The proof can be found in [Deimling (1985), Chapter 1, Sections 1 and 2].

1.4 Existence of eigenvalues

The following theorem establishes the main result of this chapter.

THEOREM 1.4.1. *Assume that the eigenvalue problem (1.1.1) is locally definite. Then, for every* $\sigma \in \{-1,1\}$ *and every multiindex (1.2.3), there exists a uniquely determined eigenvalue* $\lambda \in S^k$ *of signed index* (i,σ).

Proof. Uniqueness was shown in Theorem 1.2.3. For the existence proof, we assume that σ is positive; the negative case is analogous. We fix $u^* \in U$, and write $\lambda^* := -\delta(u^*) / \|\delta(u^*)\|$. Then $\lambda^* \in P^-$, so $P^+ \subset S^k \setminus \{\lambda^*\}$. By Lemma 1.2.1, we can choose a relatively open subset D of S^k containing P^+ such that its closure $\overline{D}$ is disjoint from P^-. In particular,

$$P \cap \partial D = \emptyset \quad . \tag{1.4.1}$$

Since we only consider the degree for maps on $\mathbb{R}^k$ but not for maps on S^k, it is necessary to work with a continuously differentiable and bijective map $h : L \to S^k \setminus \{\lambda^*\}$ having continuous inverse h^{-1} where $L \simeq \mathbb{R}^k$ is the orthocomplement of λ^* in $\mathbb{R}^{k+1}$.

We now wish to apply Theorem 1.3.1(ii) to the degree

$$\deg(g(t,h(.)), h^{-1}(D), 0) \;, \quad 0 \leq t \leq 1 \;, \tag{1.4.2}$$

where the mapping $g : [0,1] \times S^k \to \mathbb{R}^k$ has r^{th} component defined by

$$g_r(t,\lambda) := \sup\{f_r(t,\lambda,F_r) \mid F_r \subset H_r, \dim F_r = i_r\} \;, \tag{1.4.3}$$

and

$$f_r(t,\lambda,F_r) := \min\{ w_r(t^{1/2}u_r^* + (1-t)^{1/2}u_r)\lambda \mid u_r \in F_r \cap U_r, \operatorname{Re} \langle u_r,u_r^*\rangle_r = 0\}. \tag{1.4.4}$$

It should be noticed that in formula (1.4.4) the argument of w_r is a unit vector because the real part of $\langle u_r,u_r^*\rangle_r$ vanishes. Moreover, the family of functions

$$[0,1] \times S^k \ni (t,\lambda) \mapsto w_r(t^{1/2}u_r^* + (1-t)^{1/2}u_r)\lambda \quad , \quad \operatorname{Re} \langle u_r,u_r^*\rangle_r = 0 \quad ,$$

is uniformly equicontinuous which implies continuity of g.

Next, the set $h^{-1}(D)$ appearing in (1.4.2) is open and its closure $h^{-1}(\bar{D})$ is compact. Now assume that $g(t,h(x))$ vanishes for some $t \in [0,1]$ and some $x \in \partial h^{-1}(D)$. Then $g(t,\lambda) = 0$ for $\lambda = h(x) \in \partial D$ and, by (1.4.3), (1.4.4), there is a sequence $u^n \in U$ such that

$$W(u^n)\lambda \to 0 \quad \text{as} \quad n \to \infty \;.$$

The sequence u^n has an accumulation point $u \in U$ which satisfies $W(u)\lambda = 0$. It follows that $\lambda \in P$ contradicting (1.4.1).

Hence Theorem 1.3.1 (ii) shows that the degree (1.4.2) is independent of t. For $t = 1$, we have

$$g(1,\lambda) = W(u^*)\lambda \;,$$

and $W(u^*)\lambda$ vanishes if and only if λ is a multiple of λ^*. Hence $g(1,h(\cdot))$ is continuously differentiable on L and has exactly one zero x_0 satisfying $h(x_0) = -\lambda^* \in D$. The derivative of $g(1,h(.))$ at x_0 is $W(u^*)h'(x_0)$ which is an isomorphism because the range of $h'(x_0)$ is L. By formula (1.3.1), the degree (1.4.2) is nonzero for $t = 1$, and therefore is nonzero for every t.

It follows from Theorem 1.3.1 (i) that there is $\lambda^0 \in D$ satisfying $g(0,\lambda^0) = 0$. The maximum-minimum-principle (1.2.5) shows that

$$g_r(0,\lambda) = \rho_r(\lambda,i_r) \quad \text{for all} \quad \lambda \in S^k$$

because we can assume $\mathrm{Re}\, \langle u_r,u_r^*\rangle_r = 0$ in (1.2.5) if we replace u_r by $\exp(i\theta)u_r$ for a suitable real number θ. Therefore, λ^0 is an eigenvalue and has index i. Since $D \cap P^- = \emptyset$, the signum of λ^0 is positive.□

If we denote the uniquely determined eigenvalue of signed index (i,σ) by $\lambda^{(i,\sigma)}$ then we have

$$\lambda^{(i,\sigma)} = -\lambda^{(\tilde{i},-\sigma)} \quad ,$$

where i and $\tilde{i}$ are related by $i_r + \tilde{i}_r = \dim H_r + 1$ for all $r = 1,\dots,k$.

The next result shows that the eigenvalues $\lambda^{(i,\sigma)}$ depend continuously on the operators A_{rs}.

THEOREM 1.4.2. *Assume that the eigenvalue problem (1.1.1) is locally definite. Consider Hermitian operators* A^n_{rs} *on* H_r *for* $n = 1,2,\dots$, $r = 1,\dots,k$, $s = 0,\dots,k$, *such that* A^n_{rs} *converges to* A_{rs} *as* $n \to \infty$ *for each* r *and* s. *Then the eigenvalue problem*

$$\sum_{s=0}^{k} \lambda_s A^n_{rs} u_r = 0, \quad u_r \in U_r, \quad r = 1,\dots,k \quad , \tag{1.4.5}$$

is locally definite for n *sufficiently large, and the eigenvalues* $\lambda^n \in S^k$ *of (1.4.5) of fixed signed index* (i,σ) *converge to the eigenvalue* λ *of (1.1.1) of the same signed index* (i,σ).

Proof. The first part of the statement is an easy consequence of the compactness of U. For the second part, it will be sufficient to prove that λ is the only point of accumulation of the sequence λ^n because S^k is compact. Let μ be any accumulation point of the sequence λ^n and assume, without loss of generality, that λ^n converges to μ as $n \to \infty$. Then the convergence of operators

$$\sum_{s=0}^{k} \lambda_s^n A_{rs}^n \to \sum_{s=0}^{k} \mu_s A_{rs} \quad \text{as} \quad n \to \infty ,$$

shows that $\rho_r(\mu, i_r) = 0$ because of the continuous dependence of eigenvalues on Hermitian matrices. Hence μ is an eigenvalue of (1.1.1) and has index i . Similarly, it is easy to show that μ has signum σ . It follows from Theorem 1.2.3 that $\lambda = \mu$. □

The following criterion for (1.1.1) to be locally definite is a corollary of the results of this section.

THEOREM 1.4.3. *The eigenvalue problem (1.1.1) is locally definite if and only if, for every "sign vector"* $\varepsilon = (\varepsilon_1,\ldots,\varepsilon_k) \in \{-1,1\}^k$ *, there is a column vector* $\alpha = (\alpha_0,\ldots,\alpha_k) \in \mathbb{R}^{k+1}$ *such that*

$$\varepsilon_r\, w_r(u_r)\alpha > 0 \quad \text{for all} \quad u_r \in U_r,\ r = 1,\ldots,k \quad . \tag{1.4.6}$$

Proof. We first show that the condition is sufficient for local definiteness. Suppose that the equation $\beta W(u) = 0$ holds where $u \in U$ and $\beta = (\beta_1,\ldots,\beta_k) \in \mathbb{R}^k$ is a row vector. Then we choose $\varepsilon_r \in \{-1,1\}$ such that $\varepsilon_r\beta_r \geq 0$ for all r . Let α be the corresponding vector satisfying (1.4.6). Then

$$0 = \beta\, W(u)\alpha = \sum_{r=1}^{k} \beta_r\, w_r(u_r)\alpha$$

and every term under the sum sign is nonnegative. Hence $\beta_r\, w_r(u_r)\alpha = 0$ for all r From $w_r(u_r)\alpha \neq 0$ it follows that $\beta = 0$. This proves that (1.1.1) is locally definite.

Now we show that the condition is necessary. Assume that (1.1.1) is locally definite. It suffices to prove the existence of α satisfying (1.4.6) in the case $\varepsilon_r = -1,\ r = 1,\ldots,k$, because local definiteness is preserved when we replace A_{rs} by $-A_{rs}$ for certain r . By Theorem 1.4.1, there exists the eigenvalue λ of problem (1.1.1) of signed index $((1,\ldots,1),\ 1)$. Choose m such that the mth component λ_m of λ is nonzero. Then we set

$$A^n_{rs} := \begin{cases} A_{rs} + (\lambda_m /n)I_r & \text{if } s = m , \\ A_{rs} & \text{if } s \neq m , \end{cases}$$

where I_r denotes the identity operator on H_r. By Theorem 1.4.2, there exists the eigenvalue λ^n of (1.4.5) of signed index $((1,\dots,1), 1)$ for n sufficiently large, and λ^n converges to λ as $n \to \infty$. Hence we can choose n such that λ^n_m has the same sign as λ_m. From

$$0 \geq \sum_{s=0}^{k} \lambda^n_s \langle A^n_{rs} u_r, u_r \rangle_r = w_r(u_r)\lambda^n + (\lambda_m/n)\lambda^n_m$$

for all $u_r \in U_r$, $r = 1,\dots,k$, it follows that (1.4.6) holds with $\varepsilon_r = -1$ and $\alpha = \lambda^n$. □

1.5 Definite eigenvalue problems

The eigenvalue problem (1.1.1) is called *definite* if there is a (k+1)-tuple of real numbers $\mu = (\mu_0,\dots,\mu_k)$ such that the determinant of the $k+1$ by $k+1$ matrix

$$\begin{pmatrix} \mu_0 & \mu_1 & \cdots & \mu_k \\ \langle A_{10}u_1,u_1\rangle_1 & \langle A_{11}u_1,u_1\rangle_1 & \cdots & \langle A_{1k}u_1,u_1\rangle_1 \\ \vdots & \vdots & & \vdots \\ \langle A_{k0}u_k,u_k\rangle_k & \langle A_{k1}u_k,u_k\rangle_k & \cdots & \langle A_{kk}u_k,u_k\rangle_k \end{pmatrix} \tag{1.5.1}$$

is positive for all $u_r \in U_r$, $r = 1,\dots,k$. We remark that if we assume this determinant not to vanish then, by the connectedness of U, we have that this determinant has fixed sign. This sign can be made positive by changing μ to $-\mu$ if necessary.

Of course, every definite problem (1.1.1) is locally definite. A locally definite problem is definite if and only if there is a vector $\mu \in \mathbb{R}^{k+1}$ such that $\mu\,\delta(u)$ is positive for all $u \in U$, or, equivalently, such that

$$\mu\lambda > 0 \quad \text{for all } \lambda \in P^+ . \tag{1.5.2}$$

Hence definite problems are characterized by their property that the sets P^+ and

P^- can be separated by a hyperplane through zero.

We remark that the formulations of the theorems and proofs in Section 1.2 and Section 1.4 can be slightly simplified if we assume that the eigenvalue problem is definite. For instance, the signum of an eigenvalue λ is simply the sign of $\mu\lambda$ because of (1.5.2). The set D appearing in the proof of Theorem 1.4.1 can be chosen as $D = \{\lambda \in S^k \mid \mu\lambda > 0\}$.

In the rest of this section we shall show that "definiteness" and "local definiteness" are equivalent conditions for the eigenvalue problem (1.1.1) if $k = 1$ or $k = 2$ but that, in general, "local definiteness" is weaker than "definiteness" if $k \geq 3$. First we show that the locally definite problem (1.1.1) is definite in the case $k = 2$. The proof for $k = 1$ is simpler. By Theorem 1.4.3, there are linearly independent column vectors $\alpha, \beta \in \mathbb{R}^3$ such that

$$w_1(u_1)\alpha > 0, \; w_1(u_1)\beta > 0 \quad \text{for all} \quad u_1 \in U_1 \; , \tag{1.5.3}$$

$$w_2(u_2)\alpha < 0, \; w_2(u_2)\beta > 0 \quad \text{for all} \quad u_2 \in U_2 \; . \tag{1.5.4}$$

Let $\mu := \alpha \times \beta$ be the usual vector product of α and β . If we multiply the matrix (1.5.1) with $k = 2$ by the matrix (α, β, μ) from the right then we obtain the matrix

$$\begin{pmatrix} 0 & 0 & \|\mu\|^2 \\ w_1(u_1)\alpha & w_1(u_1)\beta & w_1(u_1)\mu \\ w_2(u_2)\alpha & w_2(u_2)\beta & w_2(u_2)\mu \end{pmatrix} .$$

The determinant of this matrix is positive for all $u_1 \in U_1$, $u_2 \in U_2$ because of (1.5.3) and (1.5.4). Hence the determinant of the matrix (1.5.1) is positive, too.

We now give an example of an eigenvalue problem (1.1.1) with $k = 3$ which is locally definite but not definite. We set $H_1 = H_2 = H_3 = \mathbb{C}^4$ and choose the inner products $\langle \, , \, \rangle_r$ as usual. The matrices A_{rs} are all diagonal and are uniquely determined by

$$W((e_1,e_1,e_1)) = \begin{pmatrix} 1 & 1 & 5 & -1 \\ 1 & 1 & -1 & 5 \\ 5 & -1 & 1 & 1 \end{pmatrix} ,$$

$$W((e_2,e_2,e_2)) = \begin{pmatrix} 5 & 1 & 1 & -1 \\ 1 & 5 & -1 & 1 \\ 1 & -1 & 5 & 1 \end{pmatrix} ,$$

$$W((e_3,e_3,e_3)) = \begin{pmatrix} 1 & 5 & 1 & -1 \\ 5 & 1 & -1 & 1 \\ 1 & -1 & 1 & 5 \end{pmatrix} ,$$

$$W((e_4,e_4,e_4)) = \begin{pmatrix} 1 & 1 & 1 & -5 \\ 1 & 1 & -5 & 1 \\ 1 & -5 & 1 & 1 \end{pmatrix} ,$$

where e_1,e_2,e_3,e_4 denotes the canonical basis of $\mathbb{C}^4$. Then our eigenvalue problem is locally definite because the condition of Theorem 1.4.3 is satisfied. In fact, we can choose α as one of the eight vectors $\pm e_1$, $\pm e_2$, $\pm e_3$, $\pm e_4$ for every given sign vector ε . Now we define vectors

$$\lambda^1 = \begin{pmatrix} -1 \\ -3 \\ 1 \\ 1 \end{pmatrix} , \lambda^2 = \begin{pmatrix} -1 \\ 1 \\ 1 \\ -3 \end{pmatrix} , \lambda^3 = \begin{pmatrix} -1 \\ 1 \\ -3 \\ 1 \end{pmatrix} , \lambda^4 = \begin{pmatrix} 3 \\ 1 \\ 1 \\ 1 \end{pmatrix} ,$$

which satisfy

$$W((e_n,e_n,e_n))\, \lambda^n = 0 \quad \text{for} \quad n = 1,2,3,4 \ .$$

Hence the vectors $(1/\sqrt{12}\,)\lambda^n$ lie in P and an easy calculation shows that they have all positive signum. Since $\lambda^1 + \lambda^2 + \lambda^3 + \lambda^4 = 0$, the condition (1.5.2) is violated for all μ , and therefore our eigenvalue problem is not definite.

1.6 Notes for Chapter 1

The eigenvalue problem (1.1.1) was investigated by Carmichael (1921a) for the first time. Atkinson (1972) studied definite eigenvalue problems in Chapter 7 of his book and locally definite problems in Chapter 10. Atkinson's Theorem 10.6.1 shows that the total number of pairwise nonequivalent eigenvalues of a locally definite problem, counted according to multiplicity, is equal to the dimension of the tensor product $H_1 \otimes \ldots \otimes H_k$. Thereby, two eigenvalues are called equivalent if they are complex multiples of each other. This result is weaker than the statement of Theorem 1.4.1.

Theorem 1.4.1 was proved by Binding (1984d). Binding's proof depends on Atkinson's results. Our proof of Theorem 1.4.1 based on the degree has been given by Binding and Volkmer (1986). This proof is independent of the results of Atkinson and it does not use tensor products. Another advantage of the use of the degree is that it is also possible to treat nonlinear problems by this method; see Binding (1980b).

For Theorem 1.4.2, we refer to Binding (1984d). Theorem 1.4.3 was shown by Atkinson by a different method; see [Atkinson (1972), Theorem 9.8.1]. The results of Section 1.5 can be found in Section 9.3 and Section 9.9 of the book of Atkinson (1972). Our example at the end of Section 1.5 is more explicit than that given by Atkinson.

CHAPTER 2

MULTIPARAMETER EIGENVALUE PROBLEMS FOR BOUNDED OPERATORS

2.1 Introduction

We suppose given k nonzero complex Hilbert spaces $(H_r, <\ ,\ >_r)$ with associated unit spheres U_r, $r = 1,\ldots,k$. The dimension of H_r may be finite or infinite. For each r, let A_r and A_{rs}, $s = 1,\ldots,k$, be compact Hermitian operators on H_r.

We shall study the inhomogeneous multiparameter eigenvalue problem

$$u_r = A_r u_r + \sum_{s=1}^{k} \lambda_s A_{rs} u_r,\ u_r \in U_r,\ r = 1,\ldots,k .$$

In order to bring this problem in the standard form (1.1.1) we rewrite our problem as

$$\sum_{s=0}^{k} \lambda_s A_{rs} u_r = 0,\ \lambda_0 = 1,\ u_r \in U_r,\ r = 1,\ldots,k , \qquad (2.1.1)$$

where $A_{ro} := A_r - I_r$, I_r denoting the identity operator on H_r. Our *eigenvalues* are $(k+1)$-tuples of scalars $\lambda = (1, \lambda_1,\ldots,\lambda_k)$ such that there exist vectors $u_r \in U_r$ satisfying the k equations (2.1.1).

We remark that it is not appropriate to deal with homogeneous eigenvalue problems in the case of infinite dimensional spaces H_r because then it would be difficult to index the eigenvalues with nonpositive λ_0. If the spaces H_r are finite dimensional then we should distinguish the inhomogeneous eigenvalue problem (2.1.1) from the homogeneous eigenvalue problem (1.1.1). An eigenvalue $\lambda \in S^k$ of (1.1.1) with nonzero λ_0 corresponds to the eigenvalue $\tilde{\lambda} = (1 / \lambda_0)\lambda$ of (2.1.1). However, if λ_0 is negative then, in general, λ and $\tilde{\lambda}$ will have different index and signum. The eigenvalues of (1.1.1) with $\lambda_0 = 0$ are lost when we go over from (1.1.1) to (2.1.1).

In the next section we index the eigenvalues of problem (2.1.1), and we prove the uniqueness of eigenvalues of given signed index under the hypothesis of local definiteness. As in Section 1.1, we call the eigenvalue problem (2.1.1) *locally definite* if

$$\operatorname{rank} W(u) = k \quad \text{for all} \quad u \in U := U_1 \times \dots \times U_k \quad , \tag{2.1.2}$$

where the matrix $W(u)$ is defined by (1.1.2).

The question of existence of eigenvalues of given signed index turns out to be a rather difficult one. After two preparatory sections, we shall answer the question by the important Theorem 2.5.3. Its proof uses the degree of maps and depends on the boundedness of certain subsets of $\mathbb{R}^{k+1}$ associated with the eigenvalue problem (2.1.1). This boundedness principle will be formulated and proved independently of our general setting in Section 2.4.

We remark that our central Theorem 2.5.3 is stated under a hypothesis which is somewhat stronger than local definiteness. We assume that the rank condition (2.1.2) remains valid if we replace the operators A_{ro} by the shifted operators $A_{ro} - \gamma_r I_r$ for all nonnegative γ_r, $r = 1,\dots,k$. More explicitely , we assume that

$$\operatorname{rank} \begin{pmatrix} \langle A_{1o}u_1,u_1\rangle_1-\gamma_1 & \langle A_{11}u_1,u_1\rangle_1 & \cdots & \langle A_{1k}u_1,u_1\rangle_1 \\ \vdots & \vdots & & \vdots \\ \langle A_{ko}u_k,u_k\rangle_k-\gamma_k & \langle A_{k1}u_k,u_k\rangle_k & \cdots & \langle A_{kk}u_k,u_k\rangle_k \end{pmatrix} = k \tag{2.1.3}$$

for all $u = (u_1,\dots,u_k) \in U$ and all nonnegative γ_r . If this condition holds then we say that the eigenvalue problem (2.1.1) is *locally definite in the strong sense*. If the spaces H_r are infinite dimensional then there is a close connection between local definiteness and its strong version demonstrated in Section 2.6.

In Section 2.7 we consider definite problems and, in particular, right and left definite problems. In Section 2.8 and Section 2.9 we prove a perturbation result for the eigenvalues of problem (2.1.1). References to the literature are given at the end of this chapter.

2.2 Some simple properties of eigenvalues

We consider the eigenvalue problem (2.1.1). For given $r = 1,\dots,k$ and $\lambda = (1,\lambda_1,\dots,\lambda_k) \in \mathbb{R}^{k+1}$, we form the operator

$$\sum_{s=0}^{k} \lambda_s A_{rs} = A_r - I_r + \sum_{s=1}^{k} \lambda_s A_{rs} \quad . \tag{2.2.1}$$

This operator is of the type "compact Hermitian minus identity". Hence we know that the eigenvalues of (2.2.1) are real, can accumulate only at -1 and those different from -1 have finite multiplicity. We list the eigenvalues of the operator (2.2.1) in decreasing order, according to multiplicity, as

$$\rho_r(\lambda,1) \geq \rho_r(\lambda,2) \geq \dots \quad . \tag{2.2.2}$$

If H_r is infinite dimensional and the number of eigenvalues of the operator (2.2.1) greater than -1 is finite, say equal to m, then we set $\rho_r(\lambda,n) = -1$ for $n > m$. Obviously, $\lambda = (1,\lambda_1,\dots,\lambda_k) \in \mathbb{R}^{k+1}$ is an eigenvalue of problem (2.1.1) if and only if there is a multiindex

$$i = (i_1,\dots,i_k)\ , \quad i_r = \begin{cases} 1,\dots,\dim H_r & \text{if } \dim H_r \text{ is finite ,} \\ 1,2,\dots & \text{if } \dim H_r \text{ is infinite ,} \end{cases} \tag{2.2.3}$$

such that

$$\rho_r(\lambda,i_r) = 0 \quad \text{for every } r = 1,\dots,k \quad . \tag{2.2.4}$$

If (2.2.4) holds then we say that the eigenvalue λ has *index* i. An eigenvalue can only have a finite number of indices which is called the *multiplicity* of the eigenvalue.

The maximum-minimum-principle for the eigenvalues of a compact Hermitian operator yields the representation (1.2.5) of $\rho_r(\lambda,i_r)$. If $\dim H_r = \infty$ and $\rho_r(\lambda,i_r) = -1$ then we have to replace maximum by supremum. This representation shows that $\rho_r(\cdot, i_r)$ is a continuous function because we can apply an equicontinuity argument similar to that used in the proof of Theorem 1.4.1.

LEMMA 2.2.1. *Let* $\lambda^n \in \mathbb{R}^{k+1}$ *be a sequence of eigenvalues, and let* i^n *be a sequence of pairwise distinct multiindices such that* λ^n *has index* i^n. *Then*

$$\lim_{n\to\infty} \|\lambda^n\| = \infty ,$$

where $\|\ \|$ *denotes Euclidean norm.*

Proof. Assume the contrary. Then, by taking subsequences, there are eigenvalues λ^n which have indices i^n such that λ^n converges to some $\lambda \in \mathbb{R}^{k+1}$ and i^n_ℓ converges to infinity as $n \to \infty$ for some $\ell \in \{1,\dots,k\}$. Then, for every positive integer m , the ordering (2.2.2) implies that

$$0 = \rho_\ell(\lambda^n, i_\ell^n) \le \rho_\ell(\lambda^n, m) \quad \text{for } n \text{ sufficiently large.}$$

Since $\rho_\ell(\cdot, m)$ is continuous, it follows that

$$\rho_\ell(\lambda, m) = \lim_{n\to\infty} \rho_\ell(\lambda^n, m) \ge 0 \quad \text{for all } m ,$$

which contradicts the fact that the operator (2.2.1) has only a finite number of nonnegative eigenvalues. □

In the rest of this section we assume that the eigenvalue problem (2.1.1) is locally definite. Local definiteness is equivalent to the condition

$$\delta(u) = (\delta_0(u),\dots,\delta_k(u)) \neq 0 \quad \text{for all } u \in U , \tag{2.2.5}$$

where the determinants $\delta_s(u)$ are defined as in Section 1.2. Condition (2.2.5) can also be written as

$$\operatorname{Ker} W(u) = \{\alpha\delta(u) \mid \alpha \text{ complex}\} \quad \text{for all } u \in U . \tag{2.2.6}$$

Now let $\lambda = (1,\lambda_1,\dots,\lambda_k)$ be an eigenvalue, and let $u_1,\dots,u_k$ be associated unit vectors satisfying (2.1.1). Then it follows that $W(u)\lambda = 0$, where $u = (u_1,\dots,u_k)$. Hence, by (2.2.6), λ is a complex multiple of $\delta(u)$. Since λ_0 is equal to 1 , the determinant $\delta_0(u)$ is nonzero and λ can be written as $\lambda = \alpha\delta(u)$ with $1/\alpha = \delta_0(u)$. In particular, we see that every eigenvalue of the locally definite problem (2.1.1) has real components. The sign of α is independent of those u satisfying $W(u)\lambda = 0$. This follows from Lemma 1.2.2 in the same way as in the proof of Lemma 1.2.1. The sign of α is equal to the sign of $\delta_0(u)$, and is called the *signum* of λ . We say that the eigenvalue λ is of *signed index* (i,σ) if λ has index i and signum σ . At this stage we do not define sets corresponding to P, P^+, P^- of Section 1.2. These sets will be introduced in Section 2.5.

The eigenvalue problem (2.1.1) can be related to matrix problems in the following way. Let P_r be orthoprojectors on H_r , $r = 1,\dots,k$, which have finite dimensional range spaces $\tilde{H}_r$. Then we consider the projected eigenvalue problem

$$\sum_{s=0}^{k} \lambda_s P_r A_{rs} u_r = 0, \lambda_0 = 1, u_r \in \tilde{H}_r \cap U_r, r = 1,\ldots,k . \tag{2.2.7}$$

The matrix $W(u)$ and the determinants $\delta_s(u)$ corresponding to (2.1.1) are identical to those corresponding to (2.2.7) for all $u = (u_1,\ldots,u_k)$ with $u_r \in \tilde{H}_r \cap U_r$. In particular, local definiteness of (2.1.1) implies local definiteness of (2.2.7).

LEMMA 2.2.2. *Let* λ^n, $n = 1,\ldots,m$ *, be a finite set of eigenvalues of the locally definite problem (2.1.1), and let* λ^n *be of signed index* (i^n,σ_n) *. Then there exist orthoprojectors* P_r *on* H_r, $r = 1,\ldots,k$ *, which have finite dimensional ranges* $\tilde{H}_r$ *such that* λ^n *is also an eigenvalue of problem (2.2.7) of signed index* (i^n, σ_n) *for every* $n = 1,\ldots,m$.

Proof. By the maximum-minimum-principle (1.2.5), there are linear subspaces F_r^n of H_r of dimension i_r^n such that

$$0 = \rho_r(\lambda^n,i_r^n) = \min\{w_r(u_r)\lambda^n \mid u_r \in F_r^n \cap U_r\} \quad \text{for each } r \text{ and } n . \tag{2.2.8}$$

Now let $\tilde{H}_r$ be a finite dimensional subspace of H_r large enough to contain the union of the spaces F_r^n , $n = 1,\ldots,m$, and let P_r denote the orthoprojector onto $\tilde{H}_r$. It then follows from (2.2.8) and the maximum-minimum-principle that

$$\tilde{\rho}_r(\lambda^n,i_r^n) = \rho_r(\lambda^n,i_r^n) = 0 \quad \text{for every } n = 1,\ldots,m ,$$

where the functions $\tilde{\rho}_r$ are defined with respect to problem (2.2.7). Consequently, λ^n is an eigenvalue of (2.2.7) of index i^n . Finally, it is obvious that λ^n has the same signum with respect to (2.2.7) as with respect to (2.1.1).□

We now use the preceding lemma to deduce the uniqueness of eigenvalues of given signed index from the uniqueness Theorem 1.2.3.

THEOREM 2.2.3. *The locally definite problem (2.1.1) has at most one eigenvalue of given signed index.*

Proof. Let λ^1 and λ^2 be eigenvalues of (2.1.1) of the same signed index (i,σ) . Then, by Lemma 2.2.2, there is a locally definite problem (2.2.7) with finite dimensional spaces $\widetilde{H}_r$ such that λ^1 and λ^2 are eigenvalues of this problem of signed index (i,σ). It follows from Theorem 1.2.3 that $\lambda^1/\|\lambda^1\| = \lambda^2/\|\lambda^2\|$. Hence λ^1 is equal to λ^2 because λ^1 and λ^2 have zeroth component 1.□

As a corollary of Lemma 2.2.1 and Theorem 2.2.3, we obtain the following result on the discreteness of the spectrum of a locally definite eigenvalue problem.

THEOREM 2.2.4. *The set of eigenvalues of the locally definite problem (2.1.1) has no finite point of accumulation.*

Proof. By Lemma 2.2.1 and Theorem 2.2.3, the set of eigenvalues which have positive signum cannot accumulate at a finite point. The same holds for the set of eigenvalues which have negative signum. This completes the proof.□

2.3 The joint range of forms

In this section we collect some well known results on the range of tuples of forms which we need in the sequel.

LEMMA 2.3.1. *Let* ψ_1, ψ_2 *be two Hermitian sesquilinear forms defined on a complex inner product space* $(H, \langle\,,\,\rangle)$ *with unit sphere* U. *Then their joint range*

$$\{(\psi_1(u,u), \psi_2(u,u)) \mid u \in U\}$$

is a convex subset of $\mathbb{R}^2$.

Proof. Without loss of generality, we assume that H has finite dimension. Then the mapping

$$U \ni u \mapsto f(u) := (\psi_1(u,u), \psi_2(u,u)) \in \mathbb{R}^2$$

is continuous. Now let $x,y \in U$ be such that $f(x)$ and $f(y)$ are distinct, and let π be the straight line through $f(x)$ and $f(y)$. It follows from Lemma 1.2.2, applied to a suitable linear combination of ψ_1, ψ_2 and $\langle\,,\,\rangle$, that the set $\{u \in U \mid f(u) \in \pi\}$ is arcwise connected. Hence, by continuity of f, $\pi \cap f(U)$ is arcwise connected. It follows that the set $\pi \cap f(U)$ is convex and therefore it contains the segment between $f(x)$ and $f(y)$. This proves that $f(U)$ is convex.□

Simple examples show that the above lemma cannot be generalized to more than two Hermitian forms.

In the following we shall refer to the weak topology of a Hilbert space. In particular, we shall use the fact that every sequence in the unit ball of a Hilbert space contains a subsequence converging weakly to a vector in that unit ball.

There are several equivalent definitions for a linear operator on a Hilbert space to be compact. We shall need the following one; see [Riesz and Nagy (1972), page 206].

LEMMA 2.3.2. *Let* B *be a linear operator on the Hilbert space* $(H, <,>)$. *Then* B *is compact if and only if its associated sesquilinear form* $<B\,.\,,\,.>$ *is completely continuous i. e. weak convergence of* x_n *to* x *and weak convergence of* y_n *to* y *imply*

$$<Bx_n, y_n> \;\rightarrow\; <Bx, y> \quad .$$

We remark that the above characterization of compactness of a linear operator is the original one used by Hilbert (1912). As a corollary we have

LEMMA 2.3.3. *Let* $B_1,\dots,B_k$ *be compact linear operators on the Hilbert space* $(H, <,>)$. *Then the set*

$$\{(<B_1x, x>,\dots,<B_kx, x>) \mid <x, x> \leq 1\} \tag{2.3.1}$$

is compact.

Proof. Consider a sequence

$$(<B_1x_n, x_n>,\dots,<B_kx_n, x_n>) \quad , \tag{2.3.2}$$

where x_n is a sequence in the unit ball of H . There is a subsequence y_m of x_n which converges weakly to some y in that unit ball. By Lemma 2.3.2, $<B_sy_m, y_m>$ tends to $<B_sy, y>$ as $m \to \infty$ for every $s = 1,\dots,k$. Hence the sequence (2.3.2) contains a subsequence converging to a vector of the set (2.3.1).□

2.4 A boundedness principle

In this section we shall prove a theorem on the boundedness of certain subsets of $\mathbb{R}^{k+1}$ which will be the basis of the proof of our main Theorem 2.5.3. The notations and the arguments used in this section are independent of those in the preceding sections.

We suppose given k nonempty subsets $W_1,\dots,W_k$ of $\mathbb{R}^{k+1}$ having the following two properties.

1) The images $P(W_r)$ of the sets W_r, $r = 1,\dots,k$, are convex under every projection P onto a two-dimensional linear subspace of $\mathbb{R}^{k+1}$.

2) For every $w_r \in W_r$ and every nonnegative number γ_r, $r = 1,\dots,k$, the system

$$w_1 - \gamma_1 a,\dots,w_k - \gamma_k a$$

is linearly independent, where

$$a := (1,0,\dots,0) \in \mathbb{R}^{k+1} .$$

If the eigenvalue problem (2.1.1) is locally definite in the strong sense then 1) and 2) are satisfied for the sets $W_r = \{w_r(u_r) \mid u_r \in U_r\}$, $w_r(u_r)$ denoting the rth row of the matrix (1.1.2). Condition 1) follows from Lemma 2.3.1 and condition 2) is just the definition of local definiteness in the strong sense. However, we shall not use this special form of the sets W_r.

Let us erect cones with vertex $-a$ over the sets W_r:

$$V_r := \{tw_r - (1-t)a \mid w_r \in W_r,\ 0 < t \leq 1\} . \qquad (2.4.1)$$

The vertex $-a$ is contained in V_r if and only if it is already contained in W_r. Obviously, the sets $V_1,\dots,V_k$ also have the properties 1) and 2).

LEMMA 2.4.1. *The system* $v_1,\dots,v_k$ *is linearly independent for all* $v_r \in \operatorname{co} V_r$, $r = 1,\dots,k$, *where* co *denotes convex hull.*

Proof. We prove by induction on ℓ that the system $v_1,\dots,v_k$ is linearly independent for all $v_r \in \mathrm{co}\, V_r$, $r = 1,\ldots,\ell$, and all $v_r \in V_r$, $r = \ell + 1,\ldots,k$. This is true for $\ell = 0$ because the sets V_r have property 2). We now assume that the above statement holds for $\ell - 1$ in place of ℓ . Let vectors v_r be given such that $v_r \in \mathrm{co}\, V_r$, $r = 1,\ldots,\ell$, and $v_r \in V_r$, $r = \ell + 1,\ldots,k$. We write $\mathbb{R}^{k+1}$ as a direct sum

$$\mathbb{R}^{k+1} = L \oplus N \quad , \text{ where } L := \mathrm{span}\{v_r \mid r = 1,\ldots,k,\ r \neq \ell\} \ .$$

Let P denote the projection onto N along L . By the induction hypothesis, N has dimension 2 and $P(V_\ell)$ does not contain the zero vector. It follows from property 1) for the set V_ℓ that $P(V_\ell) = \mathrm{co}\, P(V_\ell) = P(\mathrm{co}\, V_\ell)$. Hence $P(\mathrm{co}\, V_\ell)$ does not contain the zero vector, too. This shows that the system $v_1,\ldots,v_k$ is linearly independent which completes the induction. For $\ell = k$, we obtain the statement of the lemma. □

We shall use the notation

$$\delta_0(v_1,\ldots,v_k) := \det(a, v_1,\ldots,v_k)$$

for vectors v_r in $\mathbb{R}^{k+1}$.

LEMMA 2.4.2. *Let* $v_r \in \mathrm{co}\, V_r$, $r = 1,\ldots,k$, *satisfy* $\delta_0(v_1,\ldots,v_k) = 0$. *Then there are nonnegative real numbers* $\beta_1,\ldots,\beta_k$ *such that*

$$-a = \sum_{r=1}^{k} \beta_r v_r \ . \tag{2.4.2}$$

Proof. It follows from $\delta_0(v_1,\ldots,v_k) = 0$ and Lemma 2.4.1 that there are real numbers $\beta_1,\ldots,\beta_k$ satisfying (2.4.2). We show that β_ℓ is nonnegative for any given ℓ . Consider the equation

$$\sum_{r \neq \ell} \beta_r v_r + \frac{\beta_\ell}{t}\,(t\, v_\ell - (1-t)a) = -(1+\beta_\ell(\tfrac{1}{t} - 1))a \tag{2.4.3}$$

which holds for all nonzero t . For $0 < t \leq 1$, the vector $t\, v_\ell - (1-t)a$ lies in $\mathrm{co}\, V_\ell$ because of the definition (2.4.1) of V_ℓ . Hence equation (2.4.3) together with Lemma 2.4.1 gives

$$1 + \beta_\ell(\tfrac{1}{t} - 1) \neq 0 \quad \text{for all } 0 < t \leq 1 \ ,$$

which shows that β_ℓ is nonnegative. □

LEMMA 2.4.3. *Let* $v_r, w_r \in \operatorname{co} V_r$, $r = 1,\ldots,k$ *, and let* $\lambda \in \mathbb{R}^{k+1}$ *with* $a\lambda = 1$ *, where juxtaposition denotes the usual inner product. Assume that*

$$v_r\lambda \geq 0 \quad \textit{and} \quad w_r\lambda \geq 0 \quad \textit{for every} \quad r = 1,,,,k \quad .$$

Then

$$\operatorname{sign} \delta_0(v_1,\ldots,v_k) = \operatorname{sign} \delta_0(w_1,\ldots,w_k) \neq 0 \quad .$$

Proof. Assume the contrary. Then there is $t \in [0,1]$ such that

$$\delta_0(t\, v_1 + (1-t)w_1,\ldots,t\, v_k + (1-t)w_k) = 0 \quad .$$

By Lemma 2.4.2, there are nonnegative β_r satisfying

$$-a = \sum_{r=1}^{k} \beta_r(t\, v_r + (1-t)w_r) \quad .$$

Hence

$$-1 = -a\lambda = \sum_{r=1}^{k} (\beta_r\, t\, v_r\lambda + \beta_r(1-t)w_r\lambda) \geq 0 \quad ,$$

and this is the desired contradiction. □

LEMMA 2.4.4. *Let* $v_r, w_r \in \operatorname{co} V_r$, $r = 1,\ldots,k$ *, and let* $0 \neq \eta \in \mathbb{R}^{k+1}$ *with* $a\eta = 0$ *. Assume that there is a subset* M *of* $\{1,\ldots,k\}$ *such that*

$$w_r\eta < 0 < v_r\eta \qquad \textit{if} \quad r \in M \quad ,$$

and $$v_r = w_r\ ,\ v_r\eta = 0 \qquad \textit{if} \quad r \notin M \quad .$$

Further suppose that the vector a *is not a linear combination of those* v_r *with index* $r \notin M$ *. Then*

$$\operatorname{sign} \delta_0(v_1,\ldots,v_k) = -\operatorname{sign} \delta_0(w_1,\ldots,w_k) \neq 0 \quad .$$

Proof. Let x_r denote the point of intersection of the segment between w_r and v_r and the hyperplane $X = \{x \mid x\eta = 0\}$. If $r \notin M$ then $x_r = v_r = w_r$. By Lemma 2.4.1, the vectors $x_1,\ldots,x_k$ form a basis of X . Since the vector a lies in X , we can write

$$a = \sum_{r=1}^{k} \alpha_r x_r \tag{2.4.4}$$

with uniquely determined real numbers α_r . By assumption, there is $\ell \in M$ such that α_ℓ is nonzero.

We set

$$\tilde{w}_r := \begin{cases} x_r & \text{if } r \neq \ell , \\ w_r & \text{if } r = \ell , \end{cases}$$

and claim that

$$\operatorname{sign} \delta_0(w_1,\dots,w_k) = \operatorname{sign} \delta_0(\tilde{w}_1,\dots,\tilde{w}_k) \neq 0 . \tag{2.4.5}$$

Otherwise there would be $t \in [0,1]$ such that

$$\delta_0(tw_1 + (1-t)\tilde{w}_1,\dots,tw_k + (1-t)\tilde{w}_k) = 0 .$$

Hence, by Lemma 2.4.2, there are nonnegative β_r such that

$$-a = \beta_\ell w_\ell + \sum_{r\neq\ell} \beta_r(tw_r + (1-t)x_r) , \tag{2.4.6}$$

consequently,

$$0 = -a\eta = \beta_\ell w_\ell\eta + \sum_{r\in M\setminus\{\ell\}} \beta_r t w_r\eta .$$

Since $w_r\eta$ is negative for every $r \in M$, it follows that $\beta_\ell = 0$ and $t \beta_r = 0$ for every $r \in M \setminus \{\ell\}$. Hence, by (2.4.6), the vector a is a linear combination of those x_r with index r different from ℓ contradicting (2.4.4) and the choice of ℓ . This proves our claim (2.4.5).

Similarly, we can show that

$$\operatorname{sign} \delta_0(v_1,\dots,v_k) = \operatorname{sign} \delta_0(\tilde{v}_1,\dots,\tilde{v}_k) \neq 0 , \tag{2.4.7}$$

where

$$\tilde{v}_r := \begin{cases} x_r & \text{if } r \neq \ell \ , \\ v_r & \text{if } r = \ell \ . \end{cases}$$

Now consider the real-valued function f on $\mathbb{R}$ defined by

$$f(t) := \delta_o(x_1,\ldots,x_{\ell-1},tv_\ell + (1-t)w_\ell,x_{\ell+1},\ldots,x_k) \ .$$

This function is affine linear and has a zero t_o between 0 and 1 satisfying $x_\ell = t_o v_\ell + (1-t_o)w_\ell$. Hence sign $f(0) \cdot$ sign $f(1) \leq 0$ which implies the statement of the lemma because of (2.4.5) and (2.4.7).□

We are now in a position to prove our boundedness principle. This principle concerns the following two sets

$$Q^{\pm} := \{\lambda \in \mathbb{R}^k_* \mid \text{there are } w_r \in W_r \text{ and } z_r \in Z_r \text{ such that } \pm\delta_o(w_1,\ldots,w_k) \geq 0 \text{ and } z_r\lambda \leq 0 \leq w_r\lambda \text{ for } r = 1,\ldots,k\} \ ,$$

where Z_r are given subsets of W_r and

$$\mathbb{R}^k_* := \{\lambda \in \mathbb{R}^{k+1} \mid a\lambda = 1\} \ .$$

It follows from Lemma 2.4.3 that $Q^{\pm}$ can be rewritten as

$$Q^{\pm} = \{\lambda \in \mathbb{R}^k_* \mid \text{there are } v_r \in V_r \text{ and } z_r \in Z_r \text{ such that } \pm\delta_o(v_1,\ldots,v_k) \geq 0 \text{ and } z_r\lambda \leq 0 = v_r\lambda \text{ for } r = 1,\ldots,k\} \ .$$

We also remark that, by Lemma 2.4.3, the condition $\pm\delta_o(w_1,\ldots,w_k) \geq 0$ can be replaced by $\pm\delta_o(w_1,\ldots,w_k) > 0$ without altering $Q^{\pm}$. Similarly, $\pm\delta_o(v_1,\ldots,v_k) \geq 0$ can be replaced by $\pm\delta_o(v_1,\ldots,v_k) > 0$.

THEOREM 2.4.5. *Let* Z_r *be a subset of* W_r *, and assume that the sets* $V_r \cup \{-a\}$ *and* Z_r *are compact for every* $r = 1,\ldots k$ *. Then the following statements hold.*

(i) The sets Q^+ *and* Q^- *are closed and their intersection is empty.*

(ii) If

$$\delta_o(z_1,\ldots,z_k) > 0 \quad \text{for all } z_r \in Z_r \tag{2.4.8}$$

then Q^+ *is bounded. Similarly, if (2.4.8) holds with* " < 0 " *replacing* " > 0 " *then* Q^- *is bounded.*

Proof. (i) It follows from Lemma 2.4.3 that Q^+ and Q^- are disjoint. To prove that Q^+ is closed, let λ be in the closure of Q^+. Then there are sequences

$$\lambda^n \in \mathbb{R}^k_* \ , \ v^n_r \in V_r \ , \ z^n_r \in Z_r \tag{2.4.9}$$

satisfying

$$\lambda^n \to \lambda \quad \text{as} \quad n \to \infty \ ,$$

and
$$\delta_o(v^n_1,\ldots,v^n_k) > 0 \quad \text{for each} \quad n \ , \tag{2.4.10}$$

and
$$z^n_r \lambda^n \leq 0 = v^n_r \lambda^n \quad \text{for each} \quad r \quad \text{and} \quad n \ . \tag{2.4.11}$$

By taking subsequences, we can assume that

$$v^n_r \to v_r \in V_r \cup \{-a\} \ , \ z^n_r \to z_r \in Z_r \quad \text{as} \quad n \to \infty \tag{2.4.12}$$

because $V_r \cup \{-a\}$ and Z_r are compact. It follows that $\delta_o(v_1,\ldots,v_k)$ is non-negative and $z_r \lambda \leq 0 = v_r \lambda$ for every $r = 1,\ldots,k$. Since $\lambda \in \mathbb{R}^k_*$ and $v_r \lambda = 0$, the vector v_r is different from $-a$. Hence λ lies in Q^+ which proves that Q^+ is closed. Similarly, we can show that Q^- is closed.

(ii) We suppose that (2.4.8) is true but that Q^+ is unbounded. Then there are sequences (2.4.9) satisfying (2.4.10), (2.4.11) and

$$\| \lambda^n \| \to \infty \quad \text{as} \quad n \to \infty \ . \tag{2.4.13}$$

In addition we can suppose that

$$v^n_r = z^n_r \quad \text{if} \quad z^n_r \lambda^n = 0 \tag{2.4.14}$$

because, by Lemma 2.4.3, our assumptions (2.4.9), (2.4.10), (2.4.11) remain valid after replacing v^n_r by z^n_r if $z^n_r \lambda^n = 0$. By taking subsequences, we can also assume that (2.4.12) holds.

We now choose a minimal subset M of the set $\{1,\ldots,k\}$ such that $\delta_o(w_1,\ldots,w_k)$ is positive, where

$$w_r := \begin{cases} z_r & \text{if } r \in M , \\ v_r & \text{if } r \notin M . \end{cases}$$

This choice is possible because of our hypothesis (2.4.8). We note that $w_r \in V_r$ for each r . Similarly, we write

$$w_r^n := \begin{cases} z_r^n & \text{if } r \in M , \\ v_r^n & \text{if } r \notin M . \end{cases}$$

Since $\delta_o(w_1^n,\ldots,w_k^n)$ converges to the positive limit $\delta_o(w_1,\ldots,w_k)$ as $n \to \infty$, we obtain

$$\delta_o(w_1^n,\ldots,w_k^n) > 0 \quad \text{for large } n . \tag{2.4.15}$$

Hence, for large n , we can find $\mu^n , \mu \in \mathbb{R}_*^k$ solving the linear systems

$$w_r^n \mu^n = 0 \quad \text{for each } r , \tag{2.4.16}$$

and

$$w_r \mu = 0 \quad \text{for each } r . \tag{2.4.17}$$

We note that μ^n converges to μ as $n \to \infty$.

We now claim that

$$v_r \mu < 0 \quad \text{for every } r \in M . \tag{2.4.18}$$

This is trivial if $v_r = -a$ because $a\mu = 1$; so assume $v_r \in V_r$. If $v_r\mu$ is nonnegative for some $r \in M$ then Lemma 2.4.3 with $\lambda = \mu$ and (2.4.17) imply that

$$\text{sign } \delta_o(w_1,\ldots,w_{r-1},v_r,w_{r+1},\ldots,w_k) = \text{sign } \delta_o(w_1,\ldots,w_k) = 1$$

contradicting minimality of M . Thus the claim (2.4.18) is established. We deduce

$$v_r^n \mu^n < 0 \quad \text{for each } r \in M \text{ and large } n . \tag{2.4.19}$$

If we write $\eta^n := \lambda^n - \mu^n$ then $a\eta^n = 0$ and η^n is nonzero for large n because of (2.4.13) and $\mu^n \to \mu$. Now (2.4.11), (2.4.16), (2.4.19) give

$$v_r^n\, n^n > 0\ ,\ w_r^n\, n^n \le 0 \quad \text{for} \quad r \in M\ , \tag{2.4.20}$$

$$v_r^n = w_r^n\ , \quad w_r^n\, n^n = 0 \quad \text{for} \quad r \notin M\ .$$

By (2.4.14), $w_r^n\, n^n = 0$ for some $r \in M$ implies $v_r^n = w_r^n = z_r^n$ which is impossible because of (2.4.20). Hence

$$w_r^n\, n^n < 0 \quad \text{for each} \quad r \in M\ .$$

It now follows from Lemma 2.4.4 that

$$\text{sign}\ \delta_0(w_1^n,\dots,w_k^n) = -\text{sign}\ \delta_0(v_1^n,\dots,v_k^n) \neq 0 \quad \text{for large} \quad n\ .$$

This is the desired contradiction to (2.4.10). (2.4.15).

In a similar way, we can prove that Q^- is bounded if (2.4.8) holds with " < 0 " in place of " > 0 ".□

Finally, we note a refinement of the preceding result which will be needed in Section 2.9.

THEOREM 2.4.6. *Beside the sets* $Z_r \subset W_r \subset \mathbf{R}^{k+1}$ *suppose given sets* $Z_r^n \subset W_r^n \subset \mathbf{R}^{k+1}$ *for* $r = 1,\dots,k$, $n = 1,2,\dots$. *Assume that the sets* $W_1^n,\dots,W_k^n$ *as well as the sets* $W_1,\dots,W_k$ *satisfy the conditions 1) and 2) for every* n . *Further suppose that every sequence* $z_r^n \in Z_r^n$ *and* $w_r^n \in W_r^n$ *has a subsequence converging to a vector* $z_r \in Z_r$ *and* $w_r \in V_r \cup \{-a\}$, *respectively. Then, if we define the sets* $Q^{n,\pm}$ *for* Z_r^n , W_r^n *as the sets* $Q^{\pm}$ *were defined for* Z_r , W_r , *the following statements are true.*

If condition (2.4.8) holds then every sequence $\lambda^n \in Q^{n,+}$ *is bounded, and every of its accumulation points lies in* Q^+ . *Similarly, if (2.4.8) holds with " < 0 " replacing " > 0 " then every sequence* $\lambda^n \in Q^{n,-}$ *is bounded, and every of its accumulation points lies in* Q^- .

The proof of this theorem which is completely analogous to that of Theorem 2.4.5 is omitted. We should mention that our assumption that every sequence $w_r^n \in W_r^n$ has a subsequence converging to a vector $w_r \in V_r \cup \{-a\}$ implies that every sequence $v_r^n \in V_r^n \cup \{-a\}$ has a subsequence converging to a vector $v_r \in V_r \cup \{-a\}$. Thereby, the cone V_r^n is defined by (2.4.1) with W_r^n replacing W_r .

2.5 Existence and nonexistence of eigenvalues

We now return to the investigation of the eigenvalue problem (2.1.1) which is assumed to be locally definite in the strong sense. We have already mentioned that the sets $W_r = w_r(U_r)$ satisfy the conditions 1) and 2) of Section 2.4. The cone $V_r \cup \{-a\}$ defined by (2.4.1) now takes the form

$$V_r \cup \{-a\} = -a + \{(\langle A_r x_r, x_r\rangle_r, \langle A_{r1} x_r, x_r\rangle_r, \ldots, \langle A_{rk} x_r, x_r\rangle_r) \mid \langle x_r, x_r\rangle_r \le 1\} \,. \quad (2.5.1)$$

By Lemma 2.3.3, this set is compact. Hence we can apply Theorem 2.4.5 to obtain

THEOREM 2.5.1. *Let* E_r *be finite dimensional subspaces of* H_r , $r = 1,\ldots,k$, *and set* $E := E_1 \times \ldots \times E_k$. *Then the two sets*

$$Q_E^{\pm} := \{\lambda \in \mathbb{R}_*^k \mid \text{there are } u \in U \text{ and } e \in E \cap U \text{ such that } \pm\delta_0(u) \ge 0 \text{ and } W(e)\lambda \le 0 \le W(u)\lambda \}$$

have the following properties.

(i) Q_E^+ *and* Q_E^- *are closed and their intersection is empty.*

(ii) If $\delta_0(e)$ *is positive for all* $e \in E \cap U$ *then* Q_E^+ *is bounded. Similarly, if* $\delta_0(e)$ *is negative for all* $e \in E \cap U$ *then* Q_E^- *is bounded.*

Proof. Since E_r is finite dimensional, the set $Z_r := w_r(E_r \cap U_r)$ is compact. Hence our theorem follows from Theorem 2.4.5.□

The following simple remark will be useful.

LEMMA 2.5.2. *Let* λ *be an eigenvalue of problem (2.1.1) of signed index* (i,σ) , *and let* E_r *be a linear subspace of* H_r *of dimension at least* i_r, $r = 1,\ldots,k$. *Then* λ *is contained in* Q_E^+ *if and only if* σ *is positive, and* λ *is contained in* Q_E^- *if and only if* σ *is negative.*

Proof. Since λ is an eigenvalue which has signum σ , there is $u \in U$ such that $W(u)\lambda = 0$ and $\sigma\delta_0(u) > 0$. Since λ has index i , the maximum-minimum-principle (1.2.5) shows that there is $e \in E \cap U$ satisfying $W(e)\lambda \le 0$. Hence $\sigma = \pm 1$ implies $\lambda \in Q_E^{\pm}$. The converse statement follows from disjointness of Q_E^+ and Q_E^- ; see Theorem 2.5.1(i).□

Once we have proved Theorem 2.5.1, the question of existence of eigenvalues of problem (2.1.1) is easily answered by the next theorem which is sometimes called the "abstract oscillation theorem". Its proof is similar to the proof of Theorem 1.4.1 where now the sets $Q_E^{\pm}$ will play the role of the sets $P^{\pm}$ of Chapter 1.

THEOREM 2.5.3. *Assume that the eigenvalue problem (2.1.1) is locally definite in the strong sense. Then, for every* $\sigma \in \{-1,1\}$ *and every tuple* $i = (i_1,\dots,i_k)$ *of positive integers, the two following statements are equivalent.*

1) *There exists the (uniquely determined) eigenvalue of problem (2.1.1) of signed index* (i,σ) .
2) *There are linear subspaces* E_r *of* H_r *of dimension* i_r, $r = 1,\dots,k$, *such that*

$$\sigma\,\delta_0(e) > 0 \quad \textit{for all} \quad e \in E \cap U, \quad E := E_1 \times \dots \times E_k \ .$$

Proof. We suppose $\sigma = 1$; the case $\sigma = -1$ is analogous. To prove that 1) implies 2), let λ be an eigenvalue of (2.1.1) of signed index $(i,1)$. By the maximum-minimum-principle, there are linear subspaces E_r of H_r of dimension i_r such that

$$W(e)\lambda = 0 \leq W(u)\lambda \quad \text{for one} \quad e \in E \cap U \quad \text{and all} \quad u \in E \cap U \ .$$

Since $\lambda \notin Q_E^-$ by Lemma 2.5.2, it follows that $\delta_0(u)$ is positive for all $u \in E \cap U$.

We now turn to the proof that 2) implies 1). By Theorem 2.5.1, the set Q_E^+ is compact, the set Q_E^- is closed and their intersection is empty. Hence there is a bounded open subset Ω of $\mathbb{R}_*^k \simeq \mathbb{R}^k$ such that Q_E^+ is contained in Ω and Q_E^- is disjoint from the closure $\bar{\Omega}$ of Ω . In particular,

$$Q_E^+ \cap \partial\Omega = \emptyset \quad \text{and} \quad Q_E^- \cap \partial\Omega = \emptyset \ . \tag{2.5.2}$$

Now fix vectors $u_r^* \in E_r \cap U_r$, $r = 1,\dots,k$, and consider the degree

$$\deg(g(t,\cdot),\Omega,0), \quad 0 \leq t \leq 1 \ , \tag{2.5.3}$$

where the map $g : [0,1] \times \bar{\Omega} \to \mathbb{R}^k$ is defined by (1.4.3), (1.4.4). An argument similar to that used in the proof of Theorem 1.4.1 shows that g is continuous.

Suppose now that $g(t,\lambda)$ vanishes for some $t \in [0,1]$ and $\lambda \in \bar{\Omega}$. Then there is a sequence $w_r^n \in w_r(U_r)$ such that $w_r^n \lambda$ converges to zero as $n \to \infty$ for each $r = 1,\ldots,k$. Since the set $V_r \cup \{-a\}$ is compact, there is a vector $v_r \in V_r \cup \{-a\}$ such that $v_r\lambda = 0$ for each $r = 1,\ldots,k$. Obviously, v_r is different from $-a$ for each r . Moreover, it follows from $g(t,\lambda) = 0$ that $f_r(t,\lambda,E_r)$ defined by (1.4.4) is nonpositive. Hence $\lambda \in Q_E^+$ or $\lambda \in Q_E^-$ which proves that λ does not lie in the boundary of Ω because of (2.5.2). From Theorem 1.3.1 (ii) it follows that the degree (2.5.3) is independent of t .

For $t = 1$, we have $g(1,\lambda) = W(u^*)\lambda$, and this shows that the degree (2.5.3) is equal to $\operatorname{sign} \delta_0(u^*) = 1$ by assumption 2). For $t = 0$, we have $g_r(0,\lambda) = \rho_r(\lambda,i_r)$ by the maximum-minimum-principle. Thus Theorem 1.3.1 (i) proves the existence of $\lambda \in \Omega$ satisfying $\rho_r(\lambda,i_r) = 0$ for each $r = 1,\ldots,k$. Since $\lambda \in Q_E^+$, Lemma 2.5.2 shows that λ has positive signum. This completes the proof. □

It follows from the above theorem that, under local definiteness in the strong sense, existence of the eigenvalue of signed index (i,σ), $i = (i_1,\ldots,i_k)$, implies existence of all eigenvalues of signed index $(\tilde{i},\sigma)$ whenever $\tilde{i} = (\tilde{i}_1,\ldots,\tilde{i}_k)$ satisfies $\tilde{i}_r \leq i_r$, $r = 1,\ldots,k$.

It should be noticed that Theorem 2.5.3 is not a consequence of Theorem 1.4.1 in the case of finite dimensional spaces H_r . Hence Theorem 2.5.3 is also an important supplement to the theory of Chapter 1. Theorem 1.4.1 states that the locally definite problem (1.1.1) has a uniquely determined eigenvalue $\lambda^{(i,\sigma)} \in S^k$ for every signed index (i,σ) satisfying (1.2.3). Hence the statement that

$$\lambda_0^{(i,\sigma)} \quad \text{is positive}$$

is a third condition equivalent to the conditions 1) and 2) of Theorem 2.5.3 in the case of finite dimensional spaces.

We conclude this section with the remark that Theorem 2.5.3 is false under the weaker assumption that the eigenvalue problem is locally definite. For example, let $k = 1$, $H_1 = \mathbb{C}^2$ and

$$A_{1o} = \begin{pmatrix} 1 & 0 \\ 0 & 1 \end{pmatrix}, \quad A_{11} = \begin{pmatrix} 1 & 0 \\ 0 & 0 \end{pmatrix} .$$

Then the eigenvalue problem

$$\lambda_0 A_{1o}u_1 + \lambda_1 A_{11}u_1 = 0, \ \lambda_0 = 1, \ u_1 \in U_1 \ , \tag{2.5.4}$$

is locally definite and condition 2) of Theorem 2.5.3 is satisfied for the signed index $(1,1)$. However, there is no eigenvalue of (2.5.4) of signed index $(1,1)$ because $\lambda_0^{(1,1)} = 0$. Fortunately, condition (2.1.3) is violated for $\gamma_1 = 1$ so that problem (2.5.4) is not locally definite in the strong sense.

2.6 Local definiteness

In this section we shall prove a result on the relationship between local definiteness and its strong counterpart which will be applied to k-parameter Sturm-Liouville problems in Section 3.6.

It will be convenient to prove first

LEMMA 2.6.1. *Let* $(H, < , >)$ *be an infinite dimensional Hilbert space, and let* A *be a compact Hermitian operator on* H. *Then one and only one of the following three statements is true.*

1) *The operator* A *is positive or negative definite.*
2) *The operator* A *is positive or negative semidefinite and the dimension of its kernel is positive and finite.*
3) *There is a sequence* $u_1, u_2, \ldots$ *of unit vectors converging weakly to zero such that* $< A u_n, u_n > = 0$ *for each* n.

Proof. If 1) and 2) are false then A is semidefinite with infinite dimensional kernel or A is not semidefinite. In the first case statement 3) is obviously true. Hence let us assume that A is not semidefinite. Since we can replace A by $-A$ if necessary and since the dimension of H is infinite, we can assume that there is an infinite orthonormal system $x_0, x_1, x_2, \ldots$ of eigenvectors of A associated with eigenvalues $\lambda_0, \lambda_1, \lambda_2, \ldots$ such that λ_0 is negative and λ_n is nonnegative for each positive integer n . We then choose numbers α_n, β_n in the interval $[0,1]$ such that

$$\alpha_n^2\lambda_o + \beta_n^2\lambda_n = 0, \quad \alpha_n^2 + \beta_n^2 = 1 \quad ,$$

and set $u_n := \alpha_n x_o + \beta_n x_n$ for n positive. Then

$$\langle u_n, u_n\rangle = \alpha_n^2 + \beta_n^2 = 1, \ \langle A\, u_n\, , u_n\rangle = \alpha_n^2\lambda_o + \beta_n^2\lambda_n = 0 \quad .$$

Since λ_n converges to zero as $n \to \infty$, it follows that α_n converges to zero, too. Hence $u_n = \alpha_n x_o + \beta_n x_n$ converges weakly to zero because x_n converges weakly to zero. □

THEOREM 2.6.2. *Let the eigenvalue problem (2.1.1) be locally definite. For every* $r = 1,\ldots,k$ *and all real numbers* $\alpha_1,\ldots,\alpha_k$ *, assume that semidefiniteness of the operator*

$$A = \sum_{s=1}^{k} \alpha_s A_{rs} \tag{2.6.1}$$

implies that the dimension of the kernel of A *is* 0 *or* ∞ *. Then the eigenvalue problem (2.1.1) is locally definite in the strong sense.*

Proof. We first note that the dimension of H_r is infinite for each r because we can choose $\alpha_1 = \ldots = \alpha_k = 0$ in the assumption of our theorem. Hence, by Lemma 2.6.1, every operator of the form (2.6.1) has property 1) or property 3) of Lemma 2.6.1.

We now show that the vectors

$$w_1(u_1) - \gamma_1 a,\ldots,w_k(u_k) - \gamma_k a \tag{2.6.2}$$

are linearly independent for any given $u_r \in U_r$ and nonnegative γ_r , where $a = (1,0,\ldots,0) \in \mathbb{R}^{k+1}$. This is trivial if $\delta_0(u)$ is nonzero, so assume that $\delta_0(u)$ is zero. Since (2.1.1) is locally definite, the vectors $w_r := w_r(u_r)$, $r = 1,\ldots,k$, are linearly independent. Hence we can choose a vector w_0 such that $\det(w_0,w_1,\ldots,w_k) = 1$. For $\ell = 1,\ldots,k$ and every $x_\ell \in U_\ell$, the vector $w_\ell(x_\ell)$ can be represented by the basis $w_0,w_1,\ldots,w_k$ as

$$w_\ell(x_\ell) = \sum_{r=0}^{k} \chi_{\ell r}(x_\ell) w_r \quad . \tag{2.6.3}$$

The coefficients can be expressed by Cramer's rule

$$\chi_{\ell r}(x_\ell) = \det(w_0,\ldots,w_{r-1},w_\ell(x_\ell),w_{r+1},\ldots,w_k) \quad .$$

In particular, $\chi_{\ell o}(x_\ell)$ takes the form

$$\chi_{\ell o}(x_\ell) = \sum_{s=1}^{k} \delta_s(u) \langle A_{\ell s} x_\ell, x_\ell \rangle_\ell \quad , \tag{2.6.4}$$

because $\delta_o(u)$ vanishes.

We now consider the sets

$$Y_\ell := \{y_\ell \in U_\ell \mid \chi_{\ell o}(y_\ell) = 0\} \quad , \ell = 1,\ldots,k \quad ,$$

and claim that

$$\chi_{\ell\ell}(y_\ell) \neq 0 \text{ for all } y_\ell \in Y_\ell \quad . \tag{2.6.5}$$

In fact, $\chi_{\ell\ell}(y_\ell) = \chi_{\ell o}(y_\ell) = 0$ and (2.6.3) imply that $w_\ell(y_\ell)$ is a linear combination of the vectors $w_1,\ldots,w_{\ell-1},w_{\ell+1},\ldots,w_k$ contradicting local definiteness.

By Lemma 1.2.2, the sets Y_ℓ are arcwise connected. Hence it follows from continuity of $\chi_{\ell\ell}$ and (2.6.5) that $\chi_{\ell\ell}$ has constant sign on Y_ℓ. Since $\chi_{\ell r}(u_\ell) = 0$ if $r \neq \ell$ and $\chi_{\ell\ell}(u_\ell) = 1$, this sign is positive. Thus

$$\chi_{\ell\ell}(y_\ell) > 0 \text{ for all } y_\ell \in Y_\ell \quad . \tag{2.6.6}$$

The operator

$$A = \sum_{s=1}^{k} \delta_s(u) A_{\ell s}$$

is indefinite because of $\chi_{\ell o}(u_\ell) = 0$ and (2.6.4). Hence, by the first part of the proof, A has property 3) of Lemma 2.6.1. Thus there are sequences y_ℓ^n in Y_ℓ which converge weakly to zero. From Lemma 2.3.2 it then follows that $w_\ell(y_\ell^n)$ converges to $-a$ as $n \to \infty$. Hence

$$w_\ell(y_\ell^n) = \sum_{r=1}^{k} \chi_{\ell r}(y_\ell^n) w_r$$

gives

$$-a = \sum_{r=1}^{k} \beta_r w_r \quad , \tag{2.6.7}$$

where the coefficients β_r do not depend on ℓ . It follows from (2.6.6) that β_r is nonnegative for every $r = 1,\ldots,k$. Now (2.6.7) and nonnegativity of β_r and γ_r yield

$$\det(w_0, w_1-\gamma_1 a, \ldots, w_k-\gamma_k a) = 1 + \sum_{r=1}^{k} \gamma_r \beta_r \neq 0 \quad ,$$

proving the linear independence of the vectors (2.6.2).□

2.7 Definite eigenvalue problems

We call the eigenvalue problem (2.1.1) *definite* if there is a (k+1)-tuple of real numbers $\mu = (\mu_0,\ldots,\mu_k)$ such that the determinant of the matrix (1.5.1) is positive for all $u = (u_1,\ldots,u_k) \in U$. If this condition is satisfied then problem (2.1.1) is also locally definite, and the inequality

$$\mu\,\delta(u) > 0 \quad \text{for all} \quad u \in U \tag{2.7.1}$$

shows that the signum of an eigenvalue λ is equal to the sign of $\mu\lambda$.

We remark that definiteness automatically implies local definiteness in the strong sense if the spaces H_r are all infinite dimensional. This follows from an argument similar to that used in the proof of Theorem 2.6.2. In fact, since the vector w_0 can be taken as a positive multiple of μ , the inequality (2.6.5) holds for all $y_\ell \in U_\ell$.

There are two important special cases of definiteness namely right and left definiteness. The eigenvalue problem (2.1.1) is called *right definite* if it is definite with respect to $\mu = (1,0,\ldots,0)$ i. e. if

$$\delta_0(u) > 0 \quad \text{for all} \quad u \in U \ .$$

Obviously, right definiteness implies local definiteness in the strong sense. Hence the following theorem is a corollary of Theorem 2.5.3.

THEOREM 2.7.1. *Let the eigenvalue problem (2.1.1) be right definite. Then, for every multiindex* i *satisfying (2.2.3), there exists a uniquely determined eigenvalue of (2.1.1) which has index* i *. The signum of this eigenvalue is positive.*

We remark that a direct proof of this theorem would be simpler than that of Theorem 2.5.3. For instance, the set Q_E^- is empty, and Lemma 2.4.2 as well as Lemma 2.4.3 are trivial under the assumption of right definiteness.

The eigenvalue problem (2.1.1) is called *left definite* if the two following conditions are satisfied.

1) The operators A_{ro}, $r = 1,\dots,k$, are negative definite.
2) There are real numbers $\mu_1,\dots,\mu_k$ such that the cofactors of the zeroth column of the matrix (1.5.1) are negative for all $u \in U$.

More explicitely, condition 2) means that there are real numbers $\mu_1,\dots,\mu_k$ such that

$$\det \begin{pmatrix} <A_{11}u_1,u_1>_1 & \cdots & <A_{1k}u_1,u_1>_1 \\ \cdots & & \cdots \\ <A_{\ell-1,1}u_{\ell-1},u_{\ell-1}>_{\ell-1} & \cdots & <A_{\ell-1,k}u_{\ell-1},u_{\ell-1}>_{\ell-1} \\ \mu_1 & \cdots & \mu_k \\ <A_{\ell+1,1}u_{\ell+1},u_{\ell+1}>_{\ell+1} & \cdots & <A_{\ell+1,k}u_{\ell+1},u_{\ell+1}>_{\ell+1} \\ \cdots & & \cdots \\ <A_{k1}u_k,u_k>_k & \cdots & <A_{kk}u_k,u_k>_k \end{pmatrix} > 0 \qquad (2.7.2)$$

for every $\ell = 1,\dots,k$ and all $u \in U$.

LEMMA 2.7.2. *Let the eigenvalue problem (2.1.1) be left definite with respect to* $\mu_1,\dots,\mu_k$ *. Then it is definite with respect to* $\mu = (0,\mu_1,\dots,\mu_k)$ *, and it is locally definite in the strong sense.*

Proof. Consider the matrix (1.5.1) with $\mu_0 = 0$ and replace $<A_{ro}u_r,u_r>_r$ by $<A_{ro}u_r,u_r>_r - \gamma_r$ where γ_r is nonnegative. The expansion of the determinant of the resulting matrix with respect to the zeroth column shows that this determinant is positive. This proves the lemma. □

Since every left definite problem is locally definite in the strong sense, Theorem 2.5.3 can be applied to left definite problems. However, the formulation of the theorem does not simplify in this case apart from the calculation of the signum using (2.7.1). The real significance of right and left definiteness will come to light in the context of expansion theorems in Chapter 5.

The following diagram shows the mutual relations between the several definiteness conditions.

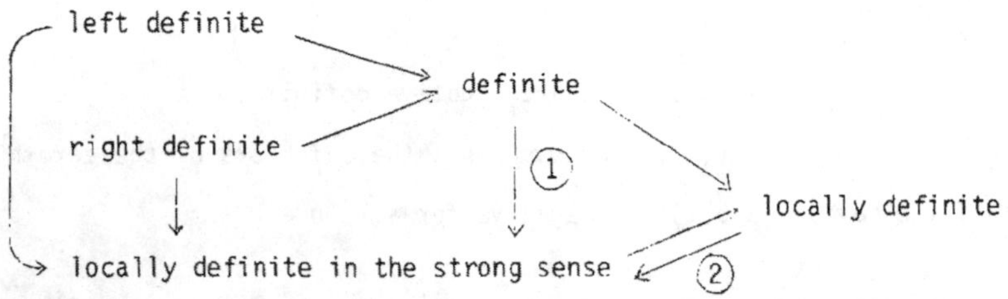

Implication ① is true if the spaces H_r are all infinite dimensional. Implication ② is true under the assumptions of Theorem 2.6.2. All other implications are true in general.

For later use, we prove a lemma concerning part 1) of the definition of left definiteness. Sometimes this condition can be achieved by use of an eigenvalue translation

$$\lambda_s = \tilde{\lambda}_s + \tau_s \quad , \quad s = 1,\dots,k \quad , \tag{2.7.3}$$

where the real numbers τ_s are suitably chosen. The translation (2.7.3) takes the eigenvalue problem (2.1.1) into

$$\sum_{s=0}^{k} \tilde{\lambda}_s \tilde{A}_{rs} u_r = 0, \quad \tilde{\lambda}_0 = 1, \; u_r \in U_r, \; r = 1,\dots,k \quad ,$$

where

$$\tilde{A}_{ro} = A_{ro} + \sum_{s=1}^{k} \tau_s A_{rs} \quad \text{for} \quad r = 1,\dots,k \quad ,$$

and $\tilde{A}_{rs} = A_{rs}$ if $s \neq 0$.

LEMMA 2.7.3. *Let the eigenvalue problem (2.1.1) be right definite. Then there exists a translation (2.7.3) which makes the operators* $\tilde{A}_{ro}$ *negative definite for every* $r = 1,\dots,k$.

Proof. Consider the eigenvalue problem

$$2\,A_r\,u_r - u_r + \sum_{s=1}^{k} \lambda_s(2\,A_{rs})\,u_r = 0, \quad u_r \in U_r,\; r = 1,\dots,k \quad ,$$

which is right definite. By Theorem 2.7.1, there exists an eigenvalue $(1,\tau_1,\dots,\tau_k)$ of this problem of index $(1,\dots,1)$. Hence the greatest eigenvalue of $2\,\widetilde{A}_{ro} + I_r$ is zero. Therefore, the greatest eigenvalue of $\widetilde{A}_{ro}$ is $-1/2$. □

The above lemma implies the following useful result. If the eigenvalue problem (2.1.1) is right definite and satisfies part 2) of the definition of left definiteness then, after a suitably chosen eigenvalue translation, we obtain a problem which is right and left definite, simultaneously.

2.8 Sequences of compact Hermitian operators

In this section we collect some results on the continuous dependence of eigenvalues on compact Hermitian operators. These results will be used in the next section and in Section 5.4.

In the following let $B_1, B_2, \dots$ and B be compact Hermitian operators on a Hilbert space $(H, \langle\ ,\ \rangle)$. Let $\rho(B,i)$ denote the ith greatest eigenvalue of B counted according to multiplicity. If this eigenvalue does not exist then we set $\rho(B,i) = 0$. The maximum-minimum-principle for the eigenvalues $\rho(B_n,i)$, $\rho(B,i)$ immediately shows that

$$|\,\rho(B_n,i) - \rho(B,i)\,| \;\le\; \|B_n - B\| \quad ,$$

where $\|\ \|$ denotes the operator norm. This proves

LEMMA 2.8.1. *If* B_n *converges to* B *with respect to the operator norm then, for every positive integer* i *,*

$$\rho(B,i) = \lim_{n\to\infty} \rho(B_n,i) \quad .$$

The assumption of the above lemma can be reformulated as follows.

LEMMA 2.8.2. *The sequence* B_n *converges to* B *with respect to the operator norm if and only if, for every sequence* $x_n \in H$ *and every* $x \in H$ *, weak convergence of* x_n *to* x *implies that* $<B_n x_n, x_n>$ *tends to* $<Bx,x>$.

Proof. Let B_n converge to B with respect to the operator norm, and let x_n converge weakly to x . Then

$$<B_n x_n, x_n> = <(B_n - B)x_n, x_n> + <Bx_n, x_n>$$

shows that $<B_n x_n, x_n>$ tends to $<Bx,x>$ because x_n is bounded and, by Lemma 2.3.2, $<Bx_n, x_n>$ tends to $<Bx,x>$.

Conversely, assume that B_n does not converge to B with respect to the operator norm. Since

$$\| B_n - B \| = \sup \{ \, | <(B_n - B)x,x> | \; | \; <x,x> \leq 1 \} ,$$

it follows that there are a positive ε , a subsequence $B_{n(m)}$ of B_n and unit vectors y_m such that

$$| <(B_{n(m)} - B)y_m, y_m> | \geq \varepsilon \quad \text{for all} \quad m . \tag{2.8.1}$$

By taking another subsequence we can additionally assume that y_m converges weakly to some $y \in H$. Since $<By_m, y_m>$ tends to $<By,y>$, (2.8.1) shows that $<B_{n(m)} y_m, y_m>$ does not tend to $<By,y>$. It follows that there is a sequence x_n converging weakly to y such that $<B_n x_n, x_n>$ does not tend to $<By,y>$.□

We use this result to prove

LEMMA 2.8.3. *Let* P_n *be a sequence of orthoprojectors on* H *such that* $P_n x$ *converges to* x *for all* $x \in H$. *Then* $P_n B P_n$ *converges to* B *with respect to the operator norm.*

Proof. Let x_n be a sequence in H converging weakly to x . Then

$$<P_n x_n, y> = <x_n, P_n y> = <x_n, y> + <x_n, P_n y - y>$$

shows that $P_n x_n$ converges weakly to x . Hence, by Lemma 2.3.2,

$$\langle P_nBP_nx_n,x_n\rangle = \langle BP_nx_n,P_nx_n\rangle \rightarrow \langle Bx,x\rangle \quad .$$

Now Lemma 2.8.2 proves that the sequence P_nBP_n converges to B with respect to the operator norm.□

If we combine Lemma 2.8.1 and Lemma 2.8.3 then we obtain, under the assumption of Lemma 2.8.3, that

$$\rho(B,i) = \lim_{n\to\infty} \rho(P_nBP_n,i) \quad .$$

2.9 Continuity of eigenvalues

Beside the eigenvalue problem (2.1.1) suppose given a sequence of eigenvalue problems

$$\sum_{s=0}^{k} \lambda_s A_{rs}^n u_r = 0 \quad , \quad \lambda_0 = 1,\ u_r \in U_r, \quad r = 1,\ldots,k \quad . \tag{2.9.1}$$

The operators A_{r0}^n are of the form $A_{r0}^n = A_r^n - I_r$, and A_r^n, A_{rs}^n are compact Hermitian operators on H_r for every $r,s = 1,\ldots,k$. We suppose that the problem (2.1.1) as well as every problem (2.9.1) is locally definite in the strong sense, and that A_r^n converges to A_r and A_{rs}^n converges to A_{rs} with respect to the operator norm for all $r,s = 1,\ldots,k$.

Under the above hypothesis we have the following result concerning the convergence of eigenvalues of (2.9.1) to those of (2.1.1).

THEOREM 2.9.1. *Let λ be an eigenvalue of the problem (2.1.1) of signed index (i,σ). Then, for sufficiently large n, there exists the eigenvalue λ^n of problem (2.9.1) of signed index (i,σ), and λ^n converges to λ as $n\to\infty$.*

Proof. We suppose $\sigma = 1$; the case $\sigma = -1$ is analogous. By Theorem 2.5.3, there are linear subspaces E_r of H_r of dimension i_r such that $\delta_0((e_1,\ldots,e_k))$ is positive for all $e_r \in E_r \cap U_r$.

We claim that $\delta_0^n((e_1,\ldots,e_k))$ is positive for all $e_r \in E_r \cap U_r$ if n is sufficiently large, where δ_0^n denotes the δ_0-determinant with respect to (2.9.1). Assuming the contrary we can find sequences $e_r^n \in E_r \cap U_r$, $r = 1,\ldots,k$, converging to some $e_r \in E_r \cap U_r$ as $n\to\infty$ such that $\delta_0^n((e_1^n,\ldots,e_k^n))$ is nonpositive. Because of the convergence of the sequences A^n_{rs} to A_{rs} , $r,s = 1,\ldots,k$, we can pass to the limit $n \to \infty$ which yields $\delta_0((e_1,\ldots,e_k)) \leq 0$. This is a contradiction which establishes our claim. Hence, by Theorem 2.5.3, the eigenvalue λ^n of signed index $(i,1)$ exists for sufficiently large n .

To prove the second part of the statement of the theorem, we appeal to Theorem 2.4.6 setting $W^n_r := w^n_r(U_r)$, $W_r = w_r(U_r)$, $Z^n_r := w^n_r(E_r \cap U_r)$, $Z_r = w_r(E_r \cap U_r)$. To verify the assumptions of that theorem, consider a sequence $w^n_r(u^n_r)$ where u^n_r is a sequence in U_r . There is a subsequence of u^n_r which converges weakly to a vector in the unit ball of H_r . It follows from Lemma 2.8.2 that the sequence $w^n_r(u^n_r)$ has a subsequence converging to an element of the cone $V_r \cup \{-a\}$; see (2.5.1). The other assumptions of Theorem 2.4.6 are also easily verified. Since λ^n is an eigenvalue of problem (2.9.1) of signed index $(i,1)$, it follows from Lemma 2.5.2 that λ^n is contained in the set $Q^{n,+}_E$ associated with problem (2.9.1). Now, by Theorem 2.4.6, the sequence λ^n is bounded, and every accumulation point of λ^n lies in Q^+_E .

For every subsequence $\lambda^{n(m)}$ of λ^n converging to some μ , the sequence of compact Hermitian operators

$$A^{n(m)}_r + \sum_{s=1}^{k} \lambda^{n(m)}_s A^{n(m)}_{rs} \tag{2.9.2}$$

converges to

$$A_r + \sum_{s=1}^{k} \mu_s A_{rs} \tag{2.9.3}$$

with respect to the operator norm. Hence, by Lemma 2.8.1, the i_rth eigenvalue of (2.9.2) which is equal to 1 converges to the i_rth eigenvalue of (2.9.3). Thus $\rho_r(\mu,i_r)$ is zero for every $r = 1,\ldots,k$. Since μ lies in Q^+_E , it follows from Lemma 2.5.2 that μ is the eigenvalue of (2.1.1) of signed index $(i,1)$. Hence the bounded sequence λ^n has the only accumulation point λ which proves the theorem.□

2.10 Notes for Chapter 2

The eigenvalue problem (2.1.1) has been investigated by Binding and Volkmer (1986) under the assumption of local definiteness. In particular, the uniqueness result of

Theorem 2.2.3 is taken from this paper.

Lemma 1.2.2 and the related convexity result of Lemma 2.3.1 were established by Hausdorff (1919).

The boundedness principle for right definite problems was proved by Binding (1984 c). A generalized version of this principle has been given by Binding and Volkmer (1986). Our presentation of the boundedness principle independent of a given eigenvalue problem is new. It would be desirable to simplify the rather technical proof of the boundedness principle but at present we have no other proof. For instance, there is no proof which yields explicit bounds for the sets $Q^{\pm}$. It is also unknown whether the boundedness principle is related to any other field of mathematics.

The existence of eigenvalues for right definite problems stated in Theorem 2.7.1 was shown by Binding (1984 c). The more general Theorem 2.5.3 was proved by Binding and Volkmer (1986) under an assumption called "A1". The condition of local definiteness in the strong sense introduced in Section 2.1 is weaker than "A1". Another advantage of our condition is its close relation to local definiteness demonstrated by Theorem 2.6.2. We remark that the proof of Theorem 2.7.1 by Binding (1984 c) depends on Theorem 1.4.1 beside the boundedness principle. Theorem 1.4.1 is applied to a sequence of eigenvalue problems for matrices which approximate the given problem (2.1.1), and the boundedness principle is needed to verify the limiting process. This method can also be used to prove Theorem 2.5.3.

Right definite problems were also studied by Atkinson (1972) in Chapter 11 of his book. Pell (1922) proved existence of eigenvalues for a left definite two-parameter problem. We refer the reader to Section 5.10 which contains more detailed notes concerning Pell's paper, and to Section 3.9 which cites the literature on eigenvalue problems for unbounded operators.

CHAPTER 3

MULTIPARAMETER EIGENVALUE PROBLEMS FOR UNBOUNDED OPERATORS

3.1 Introduction

We suppose given k nonzero complex Hilbert spaces $(H_r, < , >_r)$, $r = 1,\ldots,k$. For each r, let A_{rs}, $s = 1,\ldots,k$, be a set of k bounded Hermitian operators on H_r, and let $A_{ro} : H_r \supset D_r \to H_r$ be a selfadjoint operator, bounded above with compact resolvent.

We shall study the multiparameter eigenvalue problem

$$\sum_{s=0}^{k} \lambda_s A_{rs} x_r = 0, \quad \lambda_0 = 1, \ 0 \neq x_r \in D_r, \ r = 1,\ldots,k \ . \tag{3.1.1}$$

An eigenvalue is a (k+1)-tuple of scalars $\lambda = (1, \lambda_1,\ldots,\lambda_k)$ such that there are vectors x_r satisfying (3.1.1). If the spaces H_r are all finite dimensional then this eigenvalue problem coincides with that considered in Chapter 2, so this case will not be of interest in the present chapter.

Before we begin with our study of the eigenvalue problem (3.1.1), it will be convenient to collect some properties of the class of semibounded selfadjoint operators with compact resolvent; see Section 3.2. We shall also consider Sturm-Liouville operators as particular members of this class.

Section 3.3 shows that the eigenvalue problem (3.1.1) can be transformed into one already studied in Chapter 2. Using this transformation, it will be easy to carry over the results from Chapter 2. In particular, Theorem 2.5.3 yields an abstract oscillation theorem for the eigenvalue problem (3.1.1). In Section 3.4 we compare several definiteness conditions which can be imposed on (3.1.1).

As an important special case of the eigenvalue problem (3.1.1), we consider multiparameter boundary eigenvalue problems for linear differential operators of the second order; see Section 3.5. The abstract oscillation theorem then yields an oscillation theorem in the classical sense because there is a close connection between the indices of eigenvalues and the number of zeros of associated eigenfunctions. In Section 3.6 we prove that "local definiteness" and "local definiteness in the strong sense" are equivalent conditions for multiparameter boundary eigenvalue problems. In Sections 3.7 and 3.8 we consider two examples which are of historical interest: Klein's oscillation theorem and Stieltjes' theorem on the zeros of Lamé polynomials. More references to the literature are given in Section 3.9.

3.2 Semibounded operators with compact resolvent

In this section let $A : H \supset D \to H$ be a selfadjoint operator, bounded above with compact resolvent, defined on a dense linear subspace D of a Hilbert space $(H, \langle\ ,\ \rangle)$. The operator A has these properties if and only if it is of the form

$$A = \gamma I - T^{-1} , \tag{3.2.1}$$

where T is a compact Hermitian operator on H which is positive definite, γ is a real number and I denotes the identity operator on H. Then D is equal to the range of T. There is a uniquely determined positive definite compact operator S on H such that $S^2 = T$. Let G denote the range of S. The next lemma shows that the closure of the sesquilinear form $\langle A\cdot,\cdot\rangle$ is defined on G and can be expressed by S. We remark that A is closed as an operator because A is selfadjoint but, in general, $\langle A\cdot,\cdot\rangle$ is not closed as a form. Let us first define the closure of forms following [Kato (1966), Chapter VI, Theorem 1.17].

Let ω be a semibounded Hermitian form on a linear subspace $D(\omega)$ of H. Then ω is *closable* if

$$x_n \to 0 \quad \text{and} \quad \omega(x_n - x_m, x_n - x_m) \to 0 \quad \text{as} \quad n,m \to \infty$$

imply

$$\omega(x_n, x_n) \to 0 \quad \text{as} \quad n \to \infty$$

for every sequence $x_n \in D(\omega)$. If this condition is satisfied, ω has the *closure* $\bar{\omega}$ defined on $D(\bar{\omega})$ in the following way. $D(\bar{\omega})$ is the set of all $x \in H$ such that there exists a sequence $x_n \in D(\omega)$ with

$$x_n \to x \quad \text{and} \quad \omega(x_n - x_m, x_n - x_m) \to 0 \quad \text{as} \quad n,m \to \infty .$$

For $x,y \in D(\overline{\omega})$, $\overline{\omega}(x,y)$ is the limit

$$\overline{\omega}(x,y) := \lim_{n\to\infty} \omega(x_n,y_n) \quad ,$$

where x_n,y_n are sequences converging to x,y , respectively, according to the definition of $D(\overline{\omega})$.

LEMMA 3.2.1. *The form*

$$\omega(x,y) := \langle Ax , y\rangle \quad , \quad x,y \in D \quad ,$$

is closable. The domain $D(\overline{\omega})$ *is equal to* G *, and*

$$\overline{\omega}(x,x) = \gamma \langle x,x\rangle - \langle S^{-1}x , S^{-1}x\rangle \quad \textit{for all} \quad x \in G \ .$$

Proof. By (3.2.1) and $S^2 = T$, D is contained in G , and

$$\omega(x,x) = \gamma \langle x,x\rangle - \langle S^{-1}x , S^{-1}x\rangle \quad \text{for all} \quad x \in D \ . \tag{3.2.2}$$

Now let x_n be a sequence in D such that $x_n \to x \in H$ and $\omega(x_n-x_m,x_n-x_m) \to 0$ as $n,m \to \infty$, and write $x_n = Sy_n$. Then, by (3.2.2), y_n is a Cauchy sequence in H which converges to some $y \in H$. It follows that $x_n = Sy_n \to Sy$, and so that $x = Sy$. Hence x lies in G , and, by (3.2.2), $\omega(x_n,x_n)$ converges to $\gamma \langle x,x\rangle - \langle S^{-1}x , S^{-1}x\rangle$. In particular, this shows that ω is closable.

It remains to show that G is contained in $D(\overline{\omega})$, so let $y \in H$ and $x = Sy$. Since G is dense in H , there is a sequence $y_n \in G$ converging to y . Then the sequence $x_n = Sy_n$ lies in D and converges to x . By (3.2.2), we have $\omega(x_n-x_m , x_n-x_m) \to 0$ as $n,m \to \infty$ which proves that x lies in $D(\overline{\omega})$.□

The representation (3.2.1) of A shows that A has a discrete real spectrum bounded above i. e. the spectrum of A consists entirely of real eigenvalues of finite multiplicity which can accumulate only at $-\infty$. Therefore the eigenvalues of A can be listed in decreasing order, according to multiplicity, as $\rho(1) \geq \rho(2) \geq \ldots$. These eigenvalues satisfy the maximum-minimum-principle

$$\rho(i) = \max\{\min\{ \frac{\overline{\omega}(x,x)}{\langle x,x\rangle} \mid 0 \neq x \in F\} \mid F \subset G, \dim F = i\} \ . \tag{3.2.3}$$

Now let B be a given bounded Hermitian operator on H. Then it is known that $A + B$ is selfadjoint, bounded above with compact resolvent. The next lemma shows that the eigenvalues of this operator are related to those of $\gamma S^2 - I + SBS$. We remark that the operator $\gamma S^2 + SBS$ is compact and Hermitian.

LEMMA 3.2.2. *Let* B *be a bounded Hermitian operator on* H. *Then, for each* i, *the* ith *greatest eigenvalue* ρ *of* $A + B$ *is equal to* 0 *if and only if the* ith *greatest eigenvalue* $\tilde{\rho}$ *of* $\gamma S^2 - I + SBS$ *is equal to* 0, *where the eigenvalues are counted according to multiplicity.*

Proof. By Lemma 3.2.1, we have

$$\bar{\omega}(Sy,Sy) + \langle BSy,Sy\rangle \;=\; \langle(\gamma S^2 - I + SBS)y,y\rangle \quad , \quad y \in H \ .$$

Hence, by the maximum-minimum-principle,

$$\tilde{\rho} = \max\{\min\{ \frac{\langle(\gamma S^2 - I + SBS)y,y\rangle}{\langle y,y\rangle} \mid 0 \neq y \in E\} \mid E \subset H, \dim E = i\}$$

$$= \max\{\min\{ \frac{\bar{\omega}(x,x) + \langle Bx,x\rangle}{\langle S^{-1}x,S^{-1}x\rangle} \mid 0 \neq x \in F\} \mid F \subset G, \dim F = i\} \ .$$

Comparison with the maximum-minimum-principle (3.2.3) for ρ shows that $\tilde{\rho} = 0$ if and only if $\rho = 0$. □

It should be noted that the above lemma only states that $\tilde{\rho} = 0$ is equivalent to $\rho = 0$. In general, the eigenvalue $\tilde{\rho}$ is different from ρ.

As a special case, let us consider the linear differential operator of the second order

$$Ax \;=\; (p\,x')' + q\,x$$

subject to the boundary conditions

$$\left.\begin{aligned} \alpha_1 x(a) + \alpha_2 p(a)x'(a) + \alpha_3 x(b) + \alpha_4 p(b)x'(b) = 0 \ , \\ \beta_1 x(a) + \beta_2 p(a)x'(a) + \beta_3 x(b) + \beta_4 p(b)x'(b) = 0 \ . \end{aligned}\right\} \qquad (3.2.4)$$

It is assumed that p and q are continuous real-valued functions defined on the compact interval $[a,b]$. Moreover, p is continuously differentiable and positive on $[a,b]$. The matrix

$$\begin{pmatrix} \alpha_1 & \alpha_2 & \alpha_3 & \alpha_4 \\ \beta_1 & \beta_2 & \beta_3 & \beta_4 \end{pmatrix}$$

has real elements and is of maximal rank. Further, the boundary conditions (3.2.4) are selfadjoint i. e.

$$\det \begin{pmatrix} \alpha_1 & \alpha_2 \\ \beta_1 & \beta_2 \end{pmatrix} = \det \begin{pmatrix} \alpha_3 & \alpha_4 \\ \beta_3 & \beta_4 \end{pmatrix} \quad ; \tag{3.2.5}$$

see [Reid (1971), Chapter VI, Lemma 1.1]. We recall that Sturm-Liouville boundary conditions

$$\alpha_1 x(a) + \alpha_2 p(a)x'(a) = 0 \quad ,$$
$$\beta_3 x(b) + \beta_4 p(b)x'(b) = 0 \quad ,$$

and, if $p(a) = p(b)$, the two sets of periodic boundary conditions

$$x(a) = x(b) \quad , \quad x'(a) = x'(b) \quad ,$$

and

$$x(a) = -x(b) \; , \quad x'(a) = -x'(b) \; ,$$

are selfadjoint.

The underlying Hilbert space H is the space $L^2(a,b)$ of complex-valued square integrable functions on $[a,b]$ endowed with the usual inner product

$$\langle x,y \rangle = \int_a^b x(\xi)\overline{y(\xi)}\, d\xi \quad .$$

The domain of definition D of A consists of those differentiable functions x defined on $[a,b]$ for which x' is absolutely continuous, x'' is square integrable and x satisfies the boundary conditions (3.2.4). Then A is a selfadjoint operator, bounded above with compact resolvent; see [Dunford and Schwartz (1963), Volume II, Chapter XIII, page 1305 and Theorem 4.1].

We now wish to determine the closure of the sesquilinear form $\langle A\,.,.\rangle$. Integration by parts shows that

$$\langle Ax,y\rangle = -\int_a^b p(\xi)x'(\xi)\overline{y'(\xi)}d\xi + \int_a^b q(\xi)x(\xi)\overline{y(\xi)}d\xi + p(\xi)x'(\xi)\overline{y(\xi)}\Big|_a^b . \qquad (3.2.6)$$

Using the boundary conditions and (3.2.5), we can express the boundary form as follows

$$p(\xi)x'(\xi)\overline{y(\xi)}\Big|_a^b = \tau_1 x(a)\overline{y(a)} + \tau_2(x(a)\overline{y(b)} + x(b)\overline{y(a)}) + \tau_3 x(b)\overline{y(b)} , \qquad (3.2.7)$$

where τ_1,τ_2,τ_3 are suitably chosen real numbers; see [Reid(1971), Chapter VI, Section 1].

Let $W^1(a,b)$ denote the Hilbert space of absolutely continuous functions on $[a,b]$ which have square integrable first derivative endowed with the inner product

$$\langle x,y\rangle_W = \langle x,y\rangle + \langle x',y'\rangle .$$

For $x \in W^1(a,b)$, $a \le \xi_1,\xi_2 \le b$ and $\varepsilon > 0$, we have

$$x(\xi_2)^2 - x(\xi_1)^2 = 2\int_{\xi_1}^{\xi_2} x(\xi)x'(\xi)d\xi = 2\int_{\xi_1}^{\xi_2} \tfrac{1}{\varepsilon}x(\xi)\ \varepsilon x'(\xi)d\xi .$$

By a well known inequality, this implies that

$$|x(\xi_2)|^2 - |x(\xi_1)|^2 \le |x(\xi_2)^2 - x(\xi_1)^2| \le \frac{1}{\varepsilon^2}\int_a^b |x(\xi)|^2 d\xi + \varepsilon^2\int_a^b |x'(\xi)|^2 d\xi .$$

We integrate this inequality with respect to ξ_1 over $[a,b]$ and replace ξ_2 by ξ . This gives

$$|x(\xi)|^2 \le \left(\frac{1}{\varepsilon^2} + \frac{1}{b-a}\right)\langle x,x\rangle + \varepsilon^2\langle x',x'\rangle \quad \text{for } \xi \in [a,b] \text{ and } \varepsilon > 0 . \qquad (3.2.8)$$

It follows from (3.2.6), (3.2.7) and (3.2.8) that

$$|\langle Ax,x\rangle| \le c_1 \langle x,x\rangle_W \quad \text{for all } x \in D , \qquad (3.2.9)$$

and

$$\langle Ax,x\rangle \le c_2\langle x,x\rangle - c_3\langle x',x'\rangle \quad \text{for all } x \in D , \qquad (3.2.10)$$

where c_1,c_2,c_3 are suitably chosen positive constants. The proof of the second inequality uses (3.2.8) with ε sufficiently small and the fact that the function p attains a positive minimum on $[a,b]$; see [Reid (1971), Chapter VI, (3.7)]. We nnte that the inequality (3.2.10) shows that A is bounded above.

We now describe the closure $\overline{\omega}$ of $\omega = \langle A\,.\,.\,.\rangle$.

LEMMA 3.2.3. *The domain* $G = D(\overline{\omega})$ *is the closure of* D *with respect to the space* $W^1(a,b)$ *, and, for* $x \in G$ *, we have*

$$\overline{\omega}(x,x) = -\int_a^b p(\xi)|x'(\xi)|^2 d\xi + \int_a^b q(\xi)|x(\xi)|^2 d\xi + \tau_1|x(a)|^2 + 2\tau_2 \operatorname{Re}(x(a)\overline{x(b)}) + \tau_3|x(b)|^2 \quad , \tag{3.2.11}$$

where τ_1, τ_2, τ_3 *come from (3.2.7).*

Proof. Let $x \in G$. Then there is a sequence x_n in D such that $x_n \to x$ and $\omega(x_n - x_m, x_n - x_m) \to 0$ as $n,m \to \infty$. The inequality (3.2.10) shows that x_n is a Cauchy sequence in $W^1(a,b)$ converging to x . Hence x lies in the closure of D with respect to $W^1(a,b)$, and, by (3.2.6), $\omega(x_n, x_n)$ converges to the right-hand side of (3.2.11). Finally, the inequality (3.2.9) shows that the closure of D in $W^1(a,b)$ is a subset of G . □

We remark that D contains the space $C_0^\infty(a,b)$ of all functions on $\mathbb{R}$ which have derivatives of all orders and vanish outside a compact subset of the open interval $]a,b[$. The closure of $C_0^\infty(a,b)$ with respect to the space $W^1(a,b)$ is

$$W_0^1(a,b) = \{x \in W^1(a,b) \mid x(a) = x(b) = 0\} .$$

Hence, by Lemma 3.2.3,

$$W_0^1(a,b) \subset G \subset W^1(a,b) \quad . \tag{3.2.12}$$

For example, we have $G = W_0^1(a,b)$ under the boundary conditions $x(a) = x(b) = 0$, and $G = W^1(a,b)$ under the boundary conditions $x'(a) + \alpha\, x(a) = 0$, $x'(b) + \beta\, x(b) = 0$; compare [Kato (1966), Chapter VI, Examples 2.16, 2.17].

3.3 Transformation of eigenvalue problems

We consider the eigenvalue problem (3.1.1). Let us first introduce some notations which we already know from Chapters 1,2 but which there were related to the eigenvalue problems (1.1.1) and (2.1.1), respectively. For $\lambda = (1, \lambda_1, \ldots, \lambda_k) \in \mathbb{R}^{k+1}$ and $r = 1,\ldots,k$, we list the eigenvalues of

$$\sum_{s=0}^{k} \lambda_s A_{rs} = A_{r0} + \sum_{s=1}^{k} \lambda_s A_{rs} \tag{3.3.1}$$

in decreasing order, according to multiplicity, as

$$\rho_r(\lambda,1) \geq \rho_r(\lambda,2) \geq \dots \quad .$$

In the last section we have already remarked that the eigenvalues of the operator (3.3.1) can be listed in this way because this operator is selfadjoint, bounded above with compact resolvent. The (k+1)-tuple λ is an eigenvalue of problem (3.1.1) if and only if there is a multiindex $i = (i_1,\dots,i_k)$ such that

$$\rho_r(\lambda,i_r) = 0 \quad \text{for every} \quad r = 1,\dots,k \quad .$$

We then say that λ has *index* i. The number of indices corresponding to an eigenvalue is its *multiplicity*.

We define k by $k+1$ matrices by

$$W(x) = \begin{pmatrix} \langle A_{1o}x_1,x_1\rangle_1 & \langle A_{11}x_1,x_1\rangle_1 & \dots & \langle A_{1k}x_1,x_1\rangle_1 \\ \vdots & \vdots & & \vdots \\ \langle A_{ko}x_k,x_k\rangle_k & \langle A_{k1}x_k,x_k\rangle_k & \dots & \langle A_{kk}x_k,x_k\rangle_k \end{pmatrix} , \qquad (3.3.2)$$

where $x = (x_1,\dots,x_k)$ and $x_r \in D_r$ for each r. By Lemma 3.2.1, the sesquilinear forms

$$\omega_r(x_r,y_r) := \langle A_{ro}x_r,y_r\rangle_r \quad , \quad x_r,y_r \in D_r \quad ,$$

are closable. Using its closures $\overline{\omega}_r$, we can define the matrix $W(x)$ for all $x = (x_1,\dots,x_k)$ with $x_r \in G_r := D(\overline{\omega}_r)$, $r = 1,\dots,k$. For $s = 0,\dots,k$, we denote by $(-1)^s \delta_s(x)$ the determinant of the matrix $W(x)$ with s^{th} column deleted. We note that $\delta_0(x)$ is defined for all $x_r \in H_r$.

We call the eigenvalue problem (3.1.1) *locally definite* if

$$\operatorname{rank} W(x) = k \quad \text{for all} \quad 0 \neq x_r \in G_r \quad . \qquad (3.3.3)$$

If this condition holds then every eigenvalue λ of problem (3.1.1) is a nonzero real multiple α of some vector $\delta(x) = (\delta_0(x),\dots,\delta_k(x))$. The sign of α is equal to the sign of $\delta_0(x)$ and is called the *signum* of λ. If the eigenvalue λ has index i and signum σ then we say that λ is an eigenvalue of *signed index* (i,σ)

We now transform our problem (3.1.1) into one of the form studied in Chapter 2. We choose a real constant γ large enough for $-A_{ro}+\gamma I_r$, $r=1,\ldots,k$, to have only positive eigenvalues. As in Section 3.2, we write A_{ro} as

$$A_{ro} = \gamma I_r - S_r^{-2}, \qquad r=1,\ldots,k \quad , \tag{3.3.4}$$

where S_r is a compact, Hermitian and positive definite operator on H_r . By Lemma 3.2.1, the range of S_r is equal to $G_r = D(\bar{\omega}_r)$, and

$$\bar{\omega}_r(S_ry_r,S_ry_r) = \gamma\langle S_ry_r,S_ry_r\rangle_r - \langle y_r,y_r\rangle_r \ , \quad y_r \in H_r \ . \tag{3.3.5}$$

We set

$$\tilde{A}_{ro} := \gamma S_r^2 - I_r, \ \tilde{A}_{rs} := S_rA_{rs}S_r, \quad r,s=1,\ldots,k \ . \tag{3.3.6}$$

Then the eigenvalue problem

$$\sum_{s=0}^{k} \lambda_s \tilde{A}_{rs} u_r = 0, \quad \lambda_0 = 1, \ u_r \in U_r, \quad r=1,\ldots,k \ , \tag{3.3.7}$$

is of the type considered in Section 2.1 because the operators γS_r^2 and $\tilde{A}_{rs}$ are compact and Hermitian. As in Chapter 2, U_r denotes the unit sphere of H_r , and $U := U_1 \times \ldots \times U_k$. Here and in the sequel we use tilde to distinguish objects belonging to problem (3.3.7) from those belonging to problem (3.1.1). The relation (3.3.5) gives

$$W((S_1u_1,\ldots,S_ku_k)) = \tilde{W}(u) \quad \text{for all} \quad u=(u_1,\ldots,u_k) \in U \tag{3.3.8}$$

which implies that

$$\delta_s((S_1u_1,\ldots,S_ku_k)) = \tilde{\delta}_s(u) \quad \text{for all} \quad u \in U, \ s=0,\ldots,k \ . \tag{3.3.9}$$

(3.3.8) also shows that the problem (3.1.1) is locally definite if and only if the transformed problem (3.3.7) is locally definite. Concerning the comparison of eigenvalues of (3.1.1) and (3.3.7) we have

LEMMA 3.3.1. *Let the problem (3.1.1) be locally definite. Then* λ *is an eigenvalue of (3.1.1) of signed index* (i,σ) *if and only if* λ *is an eigenvalue of problem (3.3.7) of signed index* (i,σ) .

Proof. From Lemma 3.2.2 with

$$A := A_{ro} \quad , \quad B := \sum_{s=1}^{k} \lambda_s A_{rs}$$

it follows that $\rho_r(\lambda,i_r)$ is equal to 0 if and only if $\tilde{\rho}_r(\lambda,i_r)$ is equal to 0 for each r. Hence λ is an eigenvalue of (3.1.1) of index i if and only if λ is an eigenvalue of (3.3.7) of index i. If λ has signum σ with respect to (3.3.7) then there is $u \in U$ such that $\tilde{W}(u)\lambda = 0$ and $\sigma\,\tilde{\delta}_0(u)$ is positive. Setting $x = (S_1u_1,\ldots,S_ku_k)$, it follows from (3.3.8), (3.3.9) that $W(x)\lambda = 0$ and $\delta_0(x) = \tilde{\delta}_0(u)$. Thus λ has also signum σ with respect to problem (3.1.1).□

We now use the above lemma to carry over some of the results of Chapter 2. First, Theorem 2.2.3 proves uniqueness of an eigenvalue of the locally definite problem (3.1.1) of given signed index. Theorem 2.2.4 shows that the set of eigenvalues of the locally definite problem (3.1.1) cannot accumulate at a finite point. It is also easy to carry over Theorem 2.5.3 as follows.

THEOREM 3.3.2. *Assume that the eigenvalue problem (3.1.1) is locally definite in the strong sense i. e. assume that the rank condition (3.3.3) is true not only for the given operators* A_{ro} *but also for the shifted operators* $A_{ro} - \gamma_r I_r$ *in place of* A_{ro} *for all nonnegative* γ_r, $r = 1,\ldots,k$. *Then, for every* $\sigma \in \{-1,1\}$ *and every tuple of positive integers* $i = (i_1,\ldots,i_k)$, *the following two statements are equivalent.*

1) *There exists the (uniquely determined) eigenvalue of problem (3.1.1) of signed index* (i,σ).

2) *There are linear subspaces* F_r *of* H_r *of dimension* i_r, $r = 1,\ldots,k$, *such that*

$$\sigma\,\delta_0((x_1,\ldots,x_k)) > 0 \quad \textit{for all} \quad 0 \neq x_r \in F_r \ .$$

Proof. It follows from the assumption of the theorem that the eigenvalue problem (3.3.7) is locally definite in the strong sense. Hence the two statements of Theorem 2.5.3 corresponding to problem (3.3.7) are equivalent. By Lemma 3.3.1, the first statement of Theorem 2.5.3 is equivalent to that of Theorem 3.2.2. The second statement of Theorem 2.5.3 is equivalent to that of Theorem 3.2.2 by setting $F_r = S_r(E_r)$ and using (3.3.9) as long as F_r is contained in G_r. Therefore it remains to show that, if the second statement of our theorem holds, we can find linear subspaces F_r' of G_r of dimension i_r such that

$$\sigma\, \delta_0((x_1,\dots,x_k)) > 0 \quad \text{for all } 0 \neq x_r \in F_r' \quad .$$

This can be shown by the following arguments. Assuming 2) there is a positive ε such that $\sigma\, \delta_0(u) \geq \varepsilon$ for all $u \in F \cap U$, $F := F_1 \times \dots \times F_k$, because $F \cap U$ is compact and δ_0 is continuous. Since δ_0 is uniformly continuous on U, there is a positive τ such that $\sigma\, \delta_0(u) \geq \varepsilon/2$ for all $u = (u_1,\dots,u_k) \in U$ such that u_r is in the τ-neighborhood of the set $F_r \cap U_r$ for each r. Since G_r is dense in H_r, we can easily find linear subspaces F_r' of G_r of dimension i_r such that $F_r' \cap U_r$ is contained in the τ-neighborhood of $F_r \cap U_r$. Hence $\sigma\, \delta_0(u) \geq \varepsilon/2$ for all $u_r \in F_r' \cap U_r$. □

3.4 Definiteness conditions

For later use, we consider several definiteness conditions for the eigenvalue problem (3.1.1) and the relations between them. We have already defined local definiteness (in the strong sense) in such a way that these conditions become equivalent to the corresponding conditions of the transformed problem (3.3.7). It should be noticed that these conditions hold for (3.3.7) independently of the choice of the real number γ in (3.3.4). Theorem 2.6.2 yields

THEOREM 3.4.1. *Let the eigenvalue problem (3.1.1) be locally definite. For every* $r = 1,\dots,k$ *and all real numbers* $\alpha_1,\dots,\alpha_k$ *, assume that semidefiniteness of the operator*

$$A = \sum_{s=1}^{k} \alpha_s A_{rs}$$

implies that the dimension of $G_r \cap \operatorname{Ker} A$ *is* 0 *or* ∞ *. Then the problem (3.1.1) is locally definite in the strong sense.*

We call (3.1.1) *definite* or *right definite* or *left definite* if the transformed problem (3.3.7) has the corresponding property of Section 2.7. Then, of course, the relations between these conditions are the same as those given by the diagram in Section 2.7, where Theorem 3.4.1 replaces Theorem 2.6.2.

More explicitely, (3.1.1) is right definite if

$$\delta_0(u) > 0 \quad \text{for all} \quad u = (u_1,\dots,u_k), \quad u_r \in G_r \cap U_r \quad . \tag{3.4.1}$$

In general, it is a stronger condition to require that there is a positive ε such that

$$\delta_0(u) \geq \varepsilon \quad \text{for all} \quad u \in U \quad . \tag{3.4.2}$$

If this condition is satisfied then we call (3.1.1) *strictly right definite*. We note that strict right definiteness is only defined for the eigenvalue problem (3.1.1) but not for the problem (2.1.1) considered in Chapter 2. The reason for this is that (2.1.1) cannot satisfy (3.4.2) unless the spaces H_r are all finite dimensional because the operators A_{rs}, $r,s = 1,\dots,k$, are assumed to be compact in Chapter 2.

The problem (3.1.1) is left definite if the two following conditions hold.

1) The operators A_{r0} are negative definite for each r .
2) There are real numbers $\mu_1,\dots,\mu_k$ such that the determinants (2.7.2) are positive for all $u_r \in G_r \cap U_r$ and every $\ell = 1,\dots,k$.

The problem (3.1.1) is called *strictly left definite* if, in addition, there is a positive ε such that the determinants (2.7.2) are not smaller than ε for all $u_r \in U_r$ and every $\ell = 1,\dots,k$.

3.5 Multiparameter boundary eigenvalue problems

Let us consider the eigenvalue problem

$$(p_r x_r')' + q_r x_r + \sum_{s=1}^{k} \lambda_s a_{rs} x_r = 0, \; x_r \neq 0, \quad r = 1,\dots,k \quad , \tag{3.5.1}$$

subject to the boundary conditions

$$\left.\begin{aligned} \alpha_{r1}x_r(a_r) + \alpha_{r2}p_r(a_r)x_r'(a_r) + \alpha_{r3}x_r(b_r) + \alpha_{r4}p_r(b_r)x_r'(b_r) &= 0, \; r=1,\dots,k \,, \\ \beta_{r1}x_r(a_r) + \beta_{r2}p_r(a_r)x_r'(a_r) + \beta_{r3}x_r(b_r) + \beta_{r4}p_r(b_r)x_r'(b_r) &= 0, \; r=1,\dots,k \,. \end{aligned}\right\} \tag{3.5.2}$$

The functions p_r, q_r, a_{rs} are real-valued and continuous on the compact interval $[a_r, b_r]$. In addition, p_r is continuously differentiable and positive on $[a_r, b_r]$. The matrix $(a_{rs}(\xi_r))_{r,s=1}^{k}$ is a matrix of functions which has the characteristic feature that the functions in the rth row depend only on the rth variable $\xi_r \in [a_r, b_r]$. Matrices of this type first appeared in papers of Stäckel (1893) where he used the method of separation of variables to solve the Hamilton-Jacobi equation.

It is also assumed that the real matrix

$$\begin{pmatrix} \alpha_{r1} & \alpha_{r2} & \alpha_{r3} & \alpha_{r4} \\ \beta_{r1} & \beta_{r2} & \beta_{r3} & \beta_{r4} \end{pmatrix}$$

is of maximal rank and satisfies the selfadjointness condition (3.2.5) for each r.

An eigenvalue of problem (3.5.1), (3.5.2) is a $(k+1)$-tuple of scalars $\lambda = (1, \lambda_1, \ldots, \lambda_k)$ such that there exist nontrivial complex-valued functions x_r on $[a_r, b_r]$, twice continuously differentiable, which satisfy the differential equation (3.5.1) and the boundary conditions (3.5.2) for every $r = 1, \ldots, k$. If the components of λ are real then, of course, the functions x_r can be taken real-valued.

To satisfy the general assumptions of Section 3.1, let H_r be the Hilbert space $L^2(a_r, b_r)$ endowed with the usual inner product $\langle\ ,\ \rangle_r$. For $r,s = 1, \ldots, k$, let A_{rs} be the multiplication operator defined by

$$(A_{rs}x_r)(\xi_r) = a_{rs}(\xi_r)x_r(\xi_r) \ , \quad \xi_r \in [a_r, b_r] \ .$$

These operators A_{rs} are bounded and Hermitian on H_r. The operator A_{r0} is the differential operator

$$A_{r0}x_r = (p_r x_r')' + q_r x_r \ , \quad x_r \in D_r \ , \tag{3.5.3}$$

and D_r is defined as in Section 3.2 corresponding to the boundary conditions (3.5.2). This operator is selfadjoint, bounded above with compact resolvent. Hence the general assumptions of Section 3.1 are satisfied and we shall regard the eigenvalue problem (3.5.1), (3.5.2) as a special case of the problem (3.1.1). We note that the functions x_r, $r = 1, \ldots, k$, which satisfy the linear differential equations (3.5.1) have to be twice continuously differentiable on $[a_r, b_r]$.

For fixed r and fixed complex numbers $\lambda_1,\ldots,\lambda_k$, consider the ordinary boundary eigenvalue problem

$$(A_{r0} + \sum_{s=1}^{k} \lambda_s A_{rs})x_r = \rho_r x_r,\ 0 \neq x_r \in D_r \quad , \tag{3.5.4}$$

where ρ_r denotes the eigenvalue parameter. If the boundary conditions (3.5.2) are of Sturm-Liouville type for the given r i. e. $\alpha_{r3} = \alpha_{r4} = \beta_{r1} = \beta_{r2} = 0$ then it is clear that (3.5.4) has only simple eigenvalues. In particular, if the boundary conditions (3.5.2) are of Sturm-Liouville type for each r then our multiparameter eigenvalue problem has only eigenvalues of multiplicity one. The next theorem is another consequence of known results for the boundary eigenvalue problem (3.5.4); see [Coddington and Levinson (1955), Chapter 8, Theorems 2.1, 3.1]. It gives a description of the indices corresponding to an eigenvalue under various types of boundary conditions.

THEOREM 3.5.1. *Let* $\lambda = (1,\lambda_1,\ldots,\lambda_k) \in \mathbb{R}^{k+1}$ *be an eigenvalue of (3.5.1), (3.5.2) of index* $i = (i_1,\ldots,i_k)$ *, and let* $x_1,\ldots,x_k$ *be associated real-valued solutions of (3.5.1), (3.5.2). Then the following statements hold.*

(i) If the boundary conditions (3.5.2) are of Sturm-Liouville type for a given r *then* x_r *has exactly* $i_r - 1$ *zeros in the open interval* $]a_r,b_r[$.

(ii) If the boundary conditions (3.5.2) are of the form $x_r(a_r) = x_r(b_r)$, $x_r'(a_r) = x_r'(b_r)$ *for a given* r *then the number of zeros of* x_r *in the half-open interval* $[a_r,b_r[$ *is equal to the even integer in* $\{i_r-1, i_r\}$.

(iii) If the boundary conditions (3.5.2) are of the form $x_r(a_r) = -x_r(b_r)$, $x_r'(a_r) = -x_r'(b_r)$ *for a given* r *then the number of zeros of* x_r *in* $[a_r,b_r[$ *is equal to the odd integer in* $\{i_r-1, i_r\}$.

Our main Theorem 3.3.2 can be formulated in a very simple way for the boundary eigenvalue problem (3.5.1), (3.5.2). Let us first write the determinant $\delta_0(x)$ as

$$\delta_0(x) = \int_{\Pi} d_0(\xi)|x_1(\xi_1)|^2 \ldots |x_k(\xi_k)|^2 \, d\xi \quad , \tag{3.5.5}$$

where $x = (x_1,\ldots,x_k)$, $\xi = (\xi_1,\ldots,\xi_k) \in \Pi := [a_1,b_1] \times \ldots \times [a_k,b_k]$ and

$$d_0(\xi) = \det_{1\le r,s\le k} a_{rs}(\xi_r) \quad , \quad \xi \in \Pi \ .$$

THEOREM 3.5.2. *Let the boundary eigenvalue problem (3.5.1), (3.5.2) be locally definite. Then there is an eigenvalue of (3.5.1), (3.5.2) which has positive signum if and only if there is* $\xi \in \Pi$ *such that* $d_0(\xi)$ *is positive. If this condition is satisfied then there is a uniquely determined eigenvalue of (3.5.1), (3.5.2) of signed index* $(i,1)$ *for every tuple* $i = (i_1,\ldots,i_k)$ *of positive integers. Similar statements hold for eigenvalues having negative signum.*

Proof. If d_0 is nonpositive on Π then (3.5.5) shows that $\delta_0(x)$ is nonpositive for all $x_r \in H_r$. Hence there is no eigenvalue of problem (3.5.1), (3.5.2) which has positive signum. Let us now assume that $d_0(\xi)$ is positive for at least one $\xi \in \Pi$. Since d_0 is continuous, it follows that d_0 is positive on a product $\Omega_1 \times \ldots \times \Omega_k$ of nonempty open intervals. Then $F_r = L^2(\Omega_r)$ is an infinite dimensional subspace of H_r and, by (3.5.5), $\delta_0(x)$ is positive for all $0 \neq x_r \in F_r$. Hence statement 2) of Theorem 3.3.2 is satisfied for every multiindex $i = (i_1,\ldots,i_k)$. Thus Theorem 3.3.2 proves the theorem provided (3.5.1), (3.5.2) is locally definite in the strong sense. The next theorem shows that local definiteness and local definiteness in the strong sense are equivalent conditions for the boundary eigenvalue problem (3.5.1), (3.5.2). This completes the proof of our theorem.□

Theorem 3.5.2 together with Theorem 3.5.1 is known as the oscillation theorem for the boundary eigenvalue problem (3.5.1), (3.5.2).

Of course, there are many other multiparameter eigenvalue problems involving differential operators to which Theorem 3.3.2 can be applied. Given such a problem it remains to show that the general assumptions made for the eigenvalue problem (3.1.1) are satisfied. As an example, let us consider briefly a multiparameter eigenvalue problem involving singular Sturm-Liouville operators. In this case the closed intervals $[a_r,b_r]$ are replaced by open intervals $]a_r,b_r[$ where $-\infty \le a_r < b_r \le \infty$. On each interval $]a_r,b_r[$ there is given a continuous and positive weight function w_r . The Hilbert space H_r consists of functions on

$]a_r,b_r[$ which are square integrable with respect to the weight w_r. The singular Sturm-Liouville operators are given by

$$A_{ro}x_r = \frac{1}{w_r}((p_r x_r')' + q_r x_r) \quad ,$$

where q_r is real-valued and continuous on $]a_r,b_r[$, and p_r is positive and continuously differentiable on $]a_r,b_r[$. We assume that the equation $A_{ro}x_r = \mu x_r$ is in the limit-circle case at both endpoints of the interval $]a_r,b_r[$ for each r; see [Jörgens and Rellich (1976), Chapter III, Section 2]. We then replace the boundary conditions (3.5.2) by those appropriate in the limit-circle case; see Chapter III, Section 4 of the cited book. Subject to these boundary conditions the operator A_{ro} then leads to a selfadjoint operator on H_r bounded above with compact resolvent; see Chapter III, Section 11 of the cited book. The functions a_{rs} are supposed to be real-valued bounded and continuous on $]a_r,b_r[$ so that the corresponding multiplication operators A_{rs} are bounded and Hermitian on H_r. In this way we obtain a multiparameter eigenvalue problem which satisfies all general assumptions made for the eigenvalue problem (3.1.1). If we are in the limit-point case then, of course, we cannot expect to have eigenvalues.

3.6 Definiteness conditions for boundary eigenvalue problems

To complete the proof of Theorem 3.5.2 we need the following

THEOREM 3.6.1. *If the boundary eigenvalue problem (3.5.1), (3.5.2) is locally definite then it is also locally definite in the strong sense.*

Proof. We verify the assumption of Theorem 3.4.1. Let r and real numbers $\alpha_1,\ldots,\alpha_k$ be given such that the multiplication operator A defined by

$$(Ax_r)(\xi_r) = f(\xi_r)x_r(\xi_r) \ , \ f := \sum_{s=1}^{k} \alpha_s a_{rs} \quad ,$$

is semidefinite, say positive semidefinite. Then the continuous function f is nonnegative on $[a_r,b_r]$. If Ω denotes the set of zeros of f then $L^2(\Omega)$ is the kernel of A. If the interior of Ω is empty then $G_r \cap \operatorname{Ker} A$ contains only the function identically zero because all functions in G_r are continuous by

Lemma 3.2.3. If the interior $\mathrm{int}(\Omega)$ of Ω is nonempty then

$$G_r \cap \mathrm{Ker}\, A = G_r \cap L^2(\Omega) \supset C_0^\infty(\mathrm{int}(\Omega))$$

shows that the dimension of $G_r \cap \mathrm{Ker}\, A$ is infinite. Hence Theorem 3.4.1 can be applied proving the theorem. □

In Section 3.4 we have defind (strict) right and left definiteness of the eigenvalue problem (3.1.1). It will be convenient to characterize these conditions in the special case of the boundary eigenvalue problem (3.5.1), (3.5.2).

THEOREM 3.6.2. *The boundary eigenvalue problem (3.5.1), (3.5.2) is*

(i) strictly right definite if and only if

$$d_o(\xi) > 0 \quad \text{for all} \quad \xi \in \Pi \qquad , \tag{3.6.1}$$

(ii) right definite if and only if

$$d_o(\xi) > 0 \quad \text{on a dense subset of} \quad \Pi \quad , \tag{3.6.2}$$

(iii) strictly left definite with respect to $\mu_1,\dots,\mu_k$ *if and only if the operators* A_{ro} *are negative definite for each* r *and*

$$c_\ell(\hat{\xi}_\ell) := \det \begin{pmatrix} a_{11}(\xi_1) & \dots & a_{1k}(\xi_1) \\ \dots & & \dots \\ a_{\ell-1,1}(\xi_{\ell-1}) & \dots & a_{\ell-1,k}(\xi_{\ell-1}) \\ \mu_1 & \dots & \mu_k \\ a_{\ell+1,1}(\xi_{\ell+1}) & \dots & a_{\ell+1,k}(\xi_{\ell+1}) \\ \dots & & \dots \\ a_{k1}(\xi_k) & \dots & a_{kk}(\xi_k) \end{pmatrix} > 0 \tag{3.6.3}$$

for every $\ell = 1,\dots,k$ *and all* $\hat{\xi}_\ell = (\xi_1,\dots,\xi_{\ell-1},\xi_{\ell+1},\dots,\xi_k) \in \hat{\Pi}_\ell := \prod_{\substack{r=1 \\ r\neq\ell}}^{k} [a_r,b_r]$.

(iv) left definite with respect to $\mu_1,\dots,\mu_k$ *if and only if the operators* A_{ro} *are negative definite for each* r *and, for every* $\ell = 1,\dots,k$,

$$c_\ell(\hat{\xi}_\ell) > 0 \quad \textit{on a dense subset of } \hat{\Pi}_\ell . \tag{3.6.4}$$

Proof. (i) Assume that (3.6.1) holds. Then, since d_0 is continuous on the compact set Π, there is a positive real number ε such that $d_0(\xi) \geq \varepsilon$ for all $\xi \in \Pi$. Hence the representation (3.5.5) of $\delta_0(x)$ shows that $\delta_0(u) \geq \varepsilon$ for all $u \in U$ which proves strict right definiteness. Now assume that (3.6.1) does not hold. Then there is $\tilde{\xi} \in \Pi$ such that $d_0(\tilde{\xi})$ is nonpositive. For every given positive ε, we can find an open neighborhood $\Omega = \Omega_1 \times \ldots \times \Omega_k$ of $\tilde{\xi}$ such that $d_0(\xi) \leq \varepsilon$ for all $\xi \in \Omega$. Then we have $\delta_0(u) \leq \varepsilon$ whenever $u_r \in L^2(\Omega_r) \cap U_r$. Hence (3.5.1), (3.5.2) cannot be strictly right definite.

(ii) Assume that (3.6.2) holds. Then the integrand of (3.5.5) is nonnegative and thus $\delta_0(x)$ is nonnegative for all $x_r \in H_r$. Now let $\delta_0(x)$ vanish for some $x_r \in G_r$. Then the integrand of (3.5.5) vanishes on Π because the functions d_0 and x_r are continuous. From our assumption (3.6.2) it follows that the product $x_1(\xi_1) \ldots x_k(\xi_k)$ vanishes for all $\xi_r \in [a_r, b_r]$. Hence the function x_ℓ is identically zero for at least one ℓ. This proves right definiteness. If (3.6.2) is violated then there is a nonempty open set $\Omega = \Omega_1 \times \ldots \times \Omega_k$ such that $d_0(\xi)$ is nonpositive for all $\xi \in \Omega$. It follows that $\delta_0(x)$ is nonpositive for all $0 \neq x_r \in G_r \cap L^2(\Omega_r)$ which proves that (3.5.1), (3.5.2) is not right definite.

The proofs of (iii) and (iv) are similar.□

3.7 Klein's oscillation theorem

Klein (1881) treated the fundamental problem of potential theory for a domain whose boundary consists of confocal surfaces of the second order. He applied the method of separation of variables to the Laplace equation in ellipsoidal coordinates. This led to the problem of determining the separation constants λ_1 and λ_2 in such a way that Lamé's differential equation

$$\frac{d^2 z}{d\zeta^2} + \frac{1}{2}\left(\frac{1}{\zeta - e_1} + \frac{1}{\zeta - e_2} + \frac{1}{\zeta - e_3}\right)\frac{dz}{d\zeta} + \frac{\lambda_1 + \lambda_2 \zeta}{(\zeta - e_1)(\zeta - e_2)(\zeta - e_3)}\, z = 0 \tag{3.7.1}$$

admits two nontrivial solutions z_1 and z_2 which vanish at the endpoints of the given intervals $[a_1,b_1]$ and $[a_2,b_2]$, respectively. Thereby, it is assumed that the real numbers e_1,e_2,e_3 are fixed and satisfy $e_1 < e_2 < e_3$. In Klein's investigation the intervals $[a_1,b_1]$ and $[a_2,b_2]$ are contained in two disjoint open intervals out of the three $]e_1,e_2[$, $]e_2,e_3[$, $]e_3,\infty[$. However, at this point we only assume that the intervals $[a_1,b_1]$ and $[a_2,b_2]$ do not contain the singularities e_1,e_2,e_3 of Lamé's equation.

For abbreviation, we set

$$f(\zeta) := (\zeta-e_1)(\zeta-e_2)(\zeta-e_3) \quad ,$$
$$p(\zeta) := |f(\zeta)|^{1/2} \quad .$$

Then, if we multiply equation (3.7.1) by $p(\zeta)$, we can write our eigenvalue problem as

$$\left.\begin{aligned} &\frac{d}{d\xi_1}\left(p(\xi_1)\frac{dx_1}{d\xi_1}\right) + \frac{p(\xi_1)}{f(\xi_1)}(\lambda_1+\lambda_2\xi_1)x_1 = 0 \quad , \quad a_1 \le \xi_1 \le b_1 \ , \\ &\frac{d}{d\xi_2}\left(p(\xi_2)\frac{dx_2}{d\xi_2}\right) + \frac{p(\xi_2)}{f(\xi_2)}(\lambda_1+\lambda_2\xi_2)x_2 = 0 \quad , \quad a_2 \le \xi_2 \le b_2 \ , \end{aligned}\right\} \tag{3.7.2}$$

subject to the boundary conditions

$$x_1(a_1) = x_1(b_1) = 0 \quad , \quad x_2(a_2) = x_2(b_2) = 0 \quad . \tag{3.7.3}$$

In this form it is a special case of the boundary eigenvalue problem (3.5.1), (3.5.2). Concerning the definiteness conditions we have

LEMMA 3.7.1. *The eigenvalue problem (3.7.2), (3.7.3) is locally definite if and only if the open intervals* $]a_1,b_1[$ *and* $]a_2,b_2[$ *are disjoint. In this case the eigenvalue problem is definite with respect to* $(1,0,0)$ *or* $(-1,0,0)$ *and left definite.*

Proof. Let us assume that $]a_1,b_1[$ and $]a_2,b_2[$ are not disjoint. Then $C_0^\infty(]a_1,b_1[\cap]a_2,b_2[)$ contains a nontrivial function y. The restricted functions $x_1 = y|[a_1,b_1]$, $x_2 = y|[a_2,b_2]$ lie in D_1 and D_2, respectively, and the matrix $W((x_1,x_2))$ has identical rows violating local definiteness. Let us now assume that $]a_1,b_1[$ and $]a_2,b_2[$ are disjoint. Then the function

$$d_0((\xi_1,\xi_2)) = \frac{p(\xi_1)p(\xi_2)}{f(\xi_1)f(\xi_2)}(\xi_2-\xi_1)$$

is nonzero and of constant sign on $]a_1,b_1[\times]a_2,b_2[$. If this sign is positive then the eigenvalue problem (3.7.2), (3.7.3) is right definite by Theorem 3.6.2 (ii). Similarly, if the sign is negative then the problem is definite with respect to $(-1,0,0)$. This also proves local definiteness.

It remains to show that disjointness of $]a_1,b_1[$ and $]a_2,b_2[$ implies left definiteness. We apply Theorem 3.6.2(iv). It is clear that the operators A_{10} and A_{20} have only negative eigenvalues. To find suitable values of μ_1 and μ_2 , let us first consider the case that f is positive on $[a_1,b_1]$ and negative on $[a_2,b_2]$. Then, setting $\mu_1 = 0$, $\mu_2 = 1$, we have

$$p(\xi_2)c_1(\xi_2) = -\det\begin{pmatrix}\mu_1 & \mu_2\\ 1 & \xi_2\end{pmatrix} = 1 ,$$

$$p(\xi_1)c_2(\xi_1) = \det\begin{pmatrix}1 & \xi_1\\ \mu_1 & \mu_2\end{pmatrix} = 1 ,$$

which proves left definiteness. Next, let us assume that f is positive on $[a_1,b_1]$ and $[a_2,b_2]$. Without loss of generality, let $b_1 \le a_2$. Then, choosing $\mu_1 = 1$ and $\mu_2 \in [b_1,a_2]$, we obtain

$$p(\xi_2)c_1(\xi_2) = \det\begin{pmatrix}\mu_1 & \mu_2\\ 1 & \xi_2\end{pmatrix} = \xi_2 - \mu_2 ,$$

$$p(\xi_1)c_2(\xi_1) = \det\begin{pmatrix}1 & \xi_1\\ \mu_1 & \mu_2\end{pmatrix} = \mu_2 - \xi_1 .$$

This proves left definiteness. The remaining two distributions of signs of f on $[a_1,b_1]$ and $[a_2,b_2]$ can be handled in a similar fashion.□

We now obtain from Theorem 3.5.2 together with Theorem 3.5.1 and Lemma 3.7.1 the following

THEOREM 3.7.2. *Let the open intervals* $]a_1,b_1[$ *and* $]a_2,b_2[$ *be disjoint, and assume that the points* e_1,e_2,e_3 *are not contained in* $[a_1,b_1]$ *or* $[a_2,b_2]$. *Then, for every pair of nonnegative integers* n_1,n_2 *, there are uniquely determined* *real values of* λ_1 *and* λ_2 *such that Lamé's differential equation (3.7.1) admits a solution on* $[a_1,b_1]$ *vanishing at the endpoints and having exactly* n_1 *zeros in the interior and a second solution on* $[a_2,b_2]$ *vanishing at the endpoints and having exactly* n_2 *zeros in the interior.*

This is the content of the original oscillation theorem of Klein (1881). Klein made the stronger assumption that the intervals $[a_1,b_1]$ and $[a_2,b_2]$ are separated by e_2 or e_3 .

3.8 Stieltjes' theorem on the zeros of Lamé polynomials

In this section we want to demonstrate that the oscillation theorem of Section 3.5 is related to some results on Lamé polynomials which are treated in the theory of special functions. We have to assume that the reader has some knowledge on ordinary differential equations in the complex domain and on Jacobian elliptic functions. Everything what is needed can be found in Whittaker and Watson's book (1927) which will be cited as "WW" in the sequel.

Lamé's differential equation (3.7.1) is a Fuchsian equation having four regular singularities at e_1,e_2,e_3 and infinity. The exponents at e_1,e_2,e_3 are 0 and 1/2 and the exponents at infinity are $-n/2$ and $(n+1)/2$ where here and in the sequel we set $4\lambda_2 = -n(n+1)$, $\mathrm{Re}\, n \geq -1/2$; see [WW, Section 23.4]. It follows that, for all values of the parameters λ_1 and λ_2 and each $j = 1,2,3$, Lamé's equation has a fundamental system of solutions at $\zeta = e_j$ consisting of a function holomorphic at $\zeta = e_j$ and a function of the type $(\zeta - e_j)^{1/2}$ multiplied by a function holomorphic at $\zeta = e_j$. In general, these solutions when continued holomorphically along the real axis are not multiples of each other. If for a particular pair λ_1,λ_2 it happens that we have a nontrivial solution of Lamé's equation which is holomorphic at e_1,e_2 and e_3 simultaneously then it is an entire function because e_1,e_2,e_3 are the only singularities of Lamé's equation in the finite complex plane. Since the singularity at infinity has exponents $-n/2$ and $(n+1)/2$, every entire solution of Lamé's equation has to be a polynomial of degree $n/2$ and n is an even nonnegative integer. These polynomials are the Lamé polynomials. For further reference, we state the above result as a

LEMMA 3.8.1. *If Lamé's equation (3.7.1) has a nontrivial solution which is holomorphic at each of the three points* e_1, e_2, e_3 *then the parameter* λ_2 *is equal to* $-n(n+1)/4$ *where* n *is an even nonnegative integer and the solution is a (Lamé) polynomial of degree* $n/2$.

Concerning the existence of Lamé polynomials we have the following theorem; see [WW, Section 23.21].

THEOREM 3.8.2. *For each even nonnegative integer* n *there are precisely* $\frac{n}{2}+1$ *different real values of* λ_1 *such that Lamé's equation has a solution which is a polynomial of degree* $n/2$.

In addition to this result, Stieltjes (1885) proved the following theorem on the distribution of zeros of Lamé polynomials; see [WW, Section 23.46].

THEOREM 3.8.3. *The system of* $\frac{n}{2}+1$ *Lamé polynomials of a given degree* $n/2$ *can be arranged in order* E_n^m , $m = 0,\dots,n/2$, *in such a way that* E_n^m *has* m *(simple) zeros between* e_1 *and* e_2 *and the remaining* $\frac{n}{2} - m$ *zeros between* e_2 *and* e_3 .

In order to compare these results with those of Section 3.5 we must first formulate a suitable two-parameter eigenvalue problem for Lamé's equation. This problem is similar to that considered in the last section but now the intervals $[a_1,b_1]$ and $[a_2,b_2]$ have to be replaced by $]e_1,e_2[$, $]e_2,e_3[$, respectively. Since the endpoints of these intervals now become singularities of the equation, our eigenvalue problem does not fit in the regular setting of Section 3.5. It is of the type described at the end of Section 3.5. The weight functions are $w_1(\alpha) = w_2(\alpha) = (|\alpha-e_1||\alpha-e_2||\alpha-e_3|)^{-1/2}$. We are in the limit-circle case as can be seen from [Jörgens and Rellich (1976), Chapter III, Section 8].

However, we shall treat the eigenvalue problem by a different method which allows us to use directly the results of Section 3.5. First we simplify Lamé's equation (3.7.1) by putting $e_1 = 0$, $e_2 = 1$, $e_3 = k^{-2}$, $0 < k < 1$. There is no essential loss of generality in doing this because it can always be achieved by a linear substitution of the independent variable. Then we set $\zeta = \mathrm{sn}^2\xi$, $z(\zeta) = x(\xi)$ in Lamé's equation where sn is the Jacobian elliptic function corresponding to the modulus k ; see

[WW, Chapter XXII]. We obtain the Jacobian form of Lamé's equation [WW, Section 23.4]

$$x'' + (\mu - n(n+1)k^2 \operatorname{sn}^2\xi)x = 0 \quad , \tag{3.8.1}$$

where $\mu = 4\lambda_1 k^2$ and $4\lambda_2 = -n(n+1)$. Let K and K' denote the complete elliptic integrals belonging to the modulus k . If ξ runs from 0 to K then ζ increases from 0 to 1 . If ξ runs from K to $K+iK'$ parallel to the imaginary axis then ζ increases from 1 to k^{-2} . The point $\xi = 0$ corresponding to $\zeta = 0$ is now a regular point of (3.8.1). A solution z of (3.7.1) which is holomorphic at $\zeta = 0$ is transformed into the solution $x(\xi) = z(\operatorname{sn}^2\xi)$ of (3.8.1) satisfying $x'(0) = 0$. A solution of (3.7.1) belonging to the exponent $1/2$ at $\zeta = 0$ is transformed into a solution of (3.8.1) satisfying $x(0) = 0$. The analogous results hold at the point $\xi = K$ corresponding to $\zeta = 1$ and at the point $\xi = K+iK'$ corresponding to $\zeta = k^{-2}$. Hence the solutions of Lamé's equation (3.7.1) which are holomorphic at $e_1 = 0$, $e_2 = 1$, $e_3 = k^{-2}$ simultaneously correspond to solutions of (3.8.1) satisfying

$$x'(0) = x'(K) = x'(K+iK') = 0 \quad , \tag{3.8.2}$$

where it is understood that the solution x is holomorphic along the segments $[0,K]$ and $[K,K+iK']$.

Now we consider the two-parameter eigenvalue problem

$$\left.\begin{aligned} x_1'' + (\mu - n(n+1)k^2\operatorname{sn}^2\xi_1)x_1 &= 0 \quad , & 0 \le \xi_1 \le K \, , \\ x_2'' - (\mu - n(n+1)k^2\operatorname{sn}^2(K+i\xi_2))x_2 &= 0 \quad , & 0 \le \xi_2 \le K' , \end{aligned}\right\} \tag{3.8.3}$$

subject to the boundary conditions

$$x_1'(0) = x_1'(K) = 0 \ , \ x_2'(0) = x_2'(K') = 0 \quad . \tag{3.8.4}$$

The spectral parameters are λ_1 and λ_2 or μ and $n(n+1)$. If x_1, x_2 are nontrivial solutions of (3.8.3), (3.8.4) then we can assume that $x_1(K) = x_2(0)$ so that there exists a nontrivial holomorphic solution x of (3.8.1), (3.8.2) satisfying $x(\xi_1) = x_1(\xi_1)$ and $x(K+i\xi_2) = x_2(\xi_2)$. Similarly, a nontrivial solution of (3.8.1), (3.8.2) leads to nontrivial solutions x_1, x_2 of (3.8.3), (3.8.4). Since

$$\operatorname{sn}^2(K+i\xi_2) - \operatorname{sn}^2\xi_1 > 0 \quad \text{for} \quad 0 \le \xi_1 \le K, \ 0 \le \xi_2 \le K', \ (\xi_1,\xi_2) \ne (K,0) \ ,$$

Theorem 3.6.2 (ii) shows that the eigenvalue problem (3.8.3), (3.8.4) is right definite. Hence it follows from Theorems 3.5.1, 3.5.2 that, for each pair n_1, n_2 of nonnegative integers, there are uniquely determined real values of λ_1 and λ_2 such that there exists a solution x of (3.8.1), (3.8.2) having n_1 zeros in $]0,K[$ and n_2 zeros in $]K, K+iK'[$. Transforming back to the original equation (3.7.1) we obtain the following

THEOREM 3.8.4. *For each pair* n_1, n_2 *of nonnegative integers there are uniquely determined real values of* λ_1 *and* λ_2 *such that Lamé's equation (3.7.1) has a solution which is holomorphic at* $e_1 = 0$, $e_2 = 1$ *and* $e_3 = k^{-2}$ *having* n_1 *zeros in* $]e_1, e_2[$ *and* n_2 *zeros in* $]e_2, e_3[$.

This is the result which we get from the oscillation theorem of Section 3.5. It has been proved without using the cited results of the first part of this section.

Now let us compare the content of the above theorem with those of the Theorems 3.8.2 and 3.8.3. Theorem 3.8.4 combined with the simple Lemma 3.8.1 shows that, for each pair of nonnegative integers n_1, n_2 , there is a Lamé polynomial having n_1 zeros in $]0,1[$ and n_2 zeros in $]1,k^{-2}[$ and this Lamé polynomial is uniquely determined up to a constant factor. Hence if we assume Theorem 3.8.2 then we can prove Stieltjes' Theorem 3.8.3 by induction on the degree of Lamé polynomials. For $n = 0$ the theorem is obviously true. Then assume that the statement of the theorem holds for Lamé polynomials of degree smaller than $n/2$. It follows that a Lamé polynomial of degree $n/2$ cannot have zeros outside the intervals $]0,1[$ and $]1,k^{-2}[$ because we already know that a Lamé polynomial is uniquely determined by its numbers of zeros in $]0,1[$ and $]1,k^{-2}[$. Hence a Lamé polynomial of degree $n/2$ has all its zeros in $]0,1[$ or $]1,k^{-2}[$. We now use once again the above uniqueness result which completes the induction. Theorem 3.8.4 also proves Theorem 3.8.2 as soon as we know that a Lamé polynomial cannot have zeros outside $]e_1, e_2[\cup]e_2, e_3[$.

Besides the boundary conditions (3.8.2) we can consider 7 different conditions which we obtain by replacing one or more of the quantities $x'(0)$, $x'(1)$, $x'(k^{-2})$ by $x(0)$, $x(1)$, $x(k^{-2})$, respectively. We then obtain eigenvalue problems yielding other types of Lamé functions; see [WW, Chapter XXIII, Sections 22,23,24] .

3.9 Notes for Chapter 3

The abstract oscillation Theorem 3.3.2 was proved by Binding and Browne (1978a) for strictly right definite problems, and by Binding (1983b) for strictly left definite problems. The strictly right definite case has also been treated by Binding (1980b) using the degree of maps. The proofs of the abstract oscillation theorem in the mentioned papers are more direct than our proof which depends on the results of Chapter 2. However, those proofs use the strictness of the definiteness conditions, and it seems that they cannot be generalized to weaker conditions.

The abstract oscillation Theorem 3.3.2 was proved by Binding (1984c) for right definite problems, and by Binding and Volkmer (1986) under a definiteness condition somewhat stronger than that supposed in Theorem 3.3.2. The proofs given in these papers are similar to our proof. They both use the transformation to a problem of the form (2.1.1) and ultimately depend on the boundedness principle of Section 2.4. The mentioned transformation was also used by Binding (1983a) in the special case $\gamma = 0$.

The first oscillation theorem for multiparameter Sturm-Liouville problems is the oscillation theorem of Klein (1881) given in Section 3.7. This theorem was also treated by Bôcher (1897/98a). Bôcher (1897/98b) generalized Klein's theorem to a k-parameter problem of a special type. In the next years several authors investigated oscillation theorems for two- or three-parameter Sturm-Liouville problems under various definiteness conditions. We mention Hilb (1907a), Yoshikawa (1910) and Richardson (1912), (1912/13). For strictly right definite k-parameter problems and Sturm-Liouville boundary conditions, the oscillation theorem essentially has its origin in the textbook of Ince (1926), page 251, on ordinary differential equations. This result can also be found in Atkinson (1961) and Faierman (1969).

For left definite Sturm-Liouville problems, the oscillation theorem was proved by Faierman (1979b) in the case of two parameters, and by Sleeman (1979) and Faierman (1985) in the case of k parameters. It should be mentioned that the above authors only assume the second part of the definition of left definiteness in their versions of the oscillation theorem. Since the second part of the definition of left definiteness does not imply local definiteness, those results are not contained in Theorem 3.5.2 unless we additionally assume the first part.

Under the assumption of local definiteness, the statement of Theorem 3.5.2 seems to be new.

Oscillation theorems for two-parameter problems involving periodic boundary conditions were investigated by Howe (1971) and Faierman (1975), (1982b). We can obtain such theorems from Theorem 3.5.2 together with Theorem 3.5.1 (ii) and (iii).

Once we have a theorem on the existence of eigenvalues for a multiparameter Sturm-Liouville system the next question is to ask for the asymptotic behaviour of eigenvalues and eigenfunctions. This question has been studied in several papers, we refer to Binding and Browne (1980), (1984), Faierman (1972b), (1975a), (1977), (1979a), (1980), Turyn (1980), Schäfke and Volkmer (1988).

CHAPTER 4

MULTIPARAMETER EXPANSION THEOREMS FOR HERMITIAN MATRICES

4.1 Introduction

We suppose given k inner product spaces $(H_r, \langle\ ,\ \rangle_r)$, $r = 1,\ldots,k$, all over the complex field, nonzero, and of finite dimension. For each r, let A_{rs}, $s = 0,\ldots,k$, be a set of $k + 1$ Hermitian operators on H_r.

We shall continue our study of the eigenvalue problem of Chapter 1

$$\sum_{s=0}^{k} \lambda_s A_{rs} x_r = 0 \quad , \quad 0 \neq x_r \in H_r, \quad r = 1,\ldots,k \quad . \tag{4.1.1}$$

If $\lambda = (\lambda_0,\ldots,\lambda_k)$ is an eigenvalue and x_r, $r = 1,\ldots,k$, are associated solutions of (4.1.1) then we call the tensor

$$x_1 \otimes \ldots \otimes x_k \in H := H_1 \otimes \ldots \otimes H_k$$

a *decomposable eigenvector* belonging to the eigenvalue λ. We shall treat the question whether a given tensor in the tensor product space H can be expanded into a series of these eigenvectors.

After the preparatory Section 4.2 on tensor products we shall obtain our main Theorem 4.3.6 in Section 4.3. It states that, under the assumption of local definiteness, every tensor can be expanded into a finite series of decomposable eigenvectors. Its proof depends on the existence Theorem 1.4.1 for eigenvalues. The proof is easy under the additional assumption that all eigenvalues have multiplicity one and somewhat more complicated in the general case.

As a consequence of the expansion theorem, we obtain a result on the positive definiteness of determinants of forms in Section 4.4. Section 4.5 contains the expansion theorem for definite problems, and Section 4.6 refers to the literature.

4.2 The tensor product of finite dimensional spaces

Let H_r, $r = 1,\ldots,k$, be finite dimensional linear spaces over the complex field, and let

$$H_1 \times \ldots \times H_k \ni (x_1,\ldots,x_k) \mapsto x_1 \otimes \ldots \otimes x_k \in H := H_1 \otimes \ldots \otimes H_k$$

be the canonical map associated with their tensor product. It is assumed that the reader knows the basic properties of this multilinear map; see [Atkinson (1972), Chapters 4,5] or any textbook on multilinear algebra. We recall that the decomposable tensors

$$x_1^{m_1} \otimes \ldots \otimes x_k^{m_k} \quad , \quad m_r = 1,\ldots,\dim H_r \ ,$$

constitute a basis of H whenever $x_r^{m_r}$, $m_r = 1,\ldots,\dim H_r$, is a basis of H_r for each r . In particular, the dimension of H is equal to the product of the dimensions of the spaces H_r , $r = 1,\ldots,k$.

Let ψ_r be a sesquilinear form on H_r for every $r = 1,\ldots,k$. Then their *tensorial product* $\psi := \psi_1 \otimes \ldots \otimes \psi_k$ is the sesquilinear form on H with the characterizing property

$$\psi(x_1 \otimes \ldots \otimes x_k , y_1 \otimes \ldots \otimes y_k) = \prod_{r=1}^{k} \psi_r(x_r,y_r) \quad , \quad x_r,y_r \in H_r \ . \tag{4.2.1}$$

If ψ_{rs}, $r,s = 1,\ldots,k$, is a k by k array of sesquilinear forms on H_r then their *form determinant* is defined by

$$\psi := \det_{1\le r,s\le k} \psi_{rs} := \sum_\tau \varepsilon_\tau \psi_{1\tau(1)} \otimes \ldots \otimes \psi_{k\tau(k)} \ , \tag{4.2.2}$$

where τ runs through permutations of $1,\ldots,k$ and ε_τ is 1 or -1 if τ is even or odd, respectively. We remark that the form determinant ψ is the sesquilinear form on H which is uniquely determined by

$$\psi(x_1 \otimes \ldots \otimes x_k , y_1 \otimes \ldots \otimes y_k) = \det_{1\le r,s\le k} \psi_{rs}(x_r,y_r) \quad , \quad x_r,y_r \in H_r \ , \tag{4.2.3}$$

where an ordinary determinant of scalars appears on the right-hand side. If the forms ψ_{rs} are all Hermitian then ψ is Hermitian, too.

4.3 Completeness of eigenvectors

We consider the eigenvalue problem (4.1.1). If $\lambda = (\lambda_0,\ldots,\lambda_k)$ is an eigenvalue then we define the *eigenspace* $E(\lambda)$ associated with λ to be the tensor product

$$E(\lambda) := E_1(\lambda) \otimes \ldots \otimes E_k(\lambda) \subset H = H_1 \otimes \ldots \otimes H_k \ ,$$

where

$$E_r(\lambda) := \operatorname{Ker} \sum_{s=0}^{k} \lambda_s A_{rs} \quad , \quad r = 1,\ldots,k \quad .$$

The nonzero elements of $E(\lambda)$ are the *eigenvectors* belonging to the eigenvalue λ . We remark that the eigenspace $E(\lambda)$ is spanned by the decomposable eigenvectors belonging to λ . The dimension of $E(\lambda)$ is equal to the product of the dimensions of the spaces $E_r(\lambda)$, $r = 1,\ldots,k$. If λ has real components then the dimension of $E_r(\lambda)$ agrees with the number of integers i_r such that the eigenvalue $\rho_r(\lambda,i_r)$ is zero; see Section 1.2. Hence, in this case, the dimension of $E(\lambda)$ is equal to the number of multiindices $i = (i_1,\ldots,i_k)$ satisfying (1.2.4) which was called the multiplicity of λ .

Similarly to the definition of the determinants δ_s in Section 1.2, let $(-1)^s \varphi_s$, $s = 0,\ldots,k$, be the form determinant of the k by k array

$$\langle A_{rt} \cdot , \cdot \rangle_r \quad , \quad r = 1,\ldots,k, \quad t = 0,\ldots,k, \; t \neq s \quad . \tag{4.3.1}$$

Then $\varphi_0,\ldots,\varphi_k$ are Hermitian sesquilinear forms on H which satisfy

$$\varphi_s(u_1 \otimes \ldots \otimes u_k , u_1 \otimes \ldots \otimes u_k) = \delta_s((u_1,\ldots,u_k)) \quad \text{for all unit vectors} \quad u_r \in H_r .$$

Setting

$$\varphi(x,y) := (\varphi_0(x,y),\ldots,\varphi_k(x,y)) \quad ,$$

we have

LEMMA 4.3.1. *For all* $x \in E(\lambda)$ *and* $y \in H$, $\varphi(x,y)$ *is a complex multiple of* λ *, in particular,*

$$\lambda_s \, \varphi_t(x,y) = \lambda_t \, \varphi_s(x,y) \quad , \quad s,t = 0,\ldots,k \quad .$$

Proof. Without loss of generality, we can assume that $x = x_1 \otimes \ldots \otimes x_k$ and $y = y_1 \otimes \ldots \otimes y_k$ are decomposable tensors. Since $x \in E(\lambda)$, λ satisfies the linear system

$$\sum_{s=0}^{k} \lambda_s \langle A_{rs} x_r , y_r \rangle_r \quad , \quad r = 1,\ldots,k \quad .$$

If $\varphi(x,y)$ is zero then the statement of the lemma is trivially true. If $\varphi(x,y)$ is nonzero then the above linear system has a one dimensional space of solutions spanned by $\varphi(x,y)$. Since λ is nonzero, it follows that $\varphi(x,y)$ is a multiple of λ .□

We call tensors $x,y \in H$ *φ-orthogonal* if $\varphi(x,y) = 0$. Then Lemma 4.3.1 yields

LEMMA 4.3.2. *Eigenvectors belonging to linearly independent eigenvalues* $\lambda,\mu \in \mathbb{R}^{k+1}$ *are φ-orthogonal.*

Proof. Let $x \in E(\lambda)$, $y \in E(\mu)$. By Lemma 4.3.1, $\varphi(x,y)$ is a multiple of λ and $\varphi(y,x)$ is a multiple of μ . Since φ_S is Hermitian and μ is real, $\varphi(x,y)$ is a multiple of both λ and μ . Hence linear independence of λ,μ implies $\varphi(x,y) = 0$.□

We now turn to consequences of local definiteness. Similarly to (1.2.6), local definiteness of (4.1.1) is equivalent to

$$\varphi(x,x) \neq 0 \quad \text{for all nonzero decomposable} \quad x \in H \ . \tag{4.3.2}$$

Since the dimension of $E(\lambda)$ is equal to the multiplicity of λ , we obtain from Theorem 1.4.1 the

LEMMA 4.3.3. *Let the problem (4.1.1) be locally definite, and let* Λ *denote the set of its eigenvalues* $\lambda \in S^k$ *which have positive signum. Then*

$$\dim H = \sum_{\lambda \in \Lambda} \dim E(\lambda) \ .$$

We note that the set Λ introduced above is used to index the pairwise different eigenspaces of problem (4.1.1). In place of Λ we could use any maximal set of pairwise linear independent eigenvalues.

The previous lemma takes us some way toward the proposition that, under local definiteness, H is the direct sum of the eigenspaces $E(\lambda)$, $\lambda \in \Lambda$. To complete the proof, we have to show that every system of eigenvectors $x^\lambda \in E(\lambda)$, $\lambda \in \Lambda$, is linearly independent. To this purpose we note

LEMMA 4.3.4. *Let* x^m *be a finite system of nonzero decomposable tensors which is φ-orthogonal i. e.*

$$\varphi(x^m, x^n) = 0 \quad \textit{if} \quad m \neq n \ .$$

Then, if (4.1.1) is locally definite, the system x^m *is linearly independent.*

Proof. Assume that

$$\sum_m \alpha_m x^m = 0 \ , \quad \alpha_m \ \text{complex.}$$

Then, by the φ-orthogonality of the system x^m,

$$0 = \varphi(\sum_m \alpha_m x^m, x^n) = \alpha_n \varphi(x^n, x^n) \ .$$

By (4.3.2), $\varphi(x^n, x^n)$ is nonzero. Hence α_n is zero for each n which proves the lemma.□

In general, the statement of the above lemma is false if the tensors x^m are not decomposable. In fact, if the problem (4.1.1) is not definite then there is a nonzero tensor which is φ-orthogonal to itself; see Theorem 4.5.2.

If all eigenvalues of a locally definite problem (4.1.1) have multiplicity one then the Lemmas 4.3.2, 4.3.3, 4.3.4 show that H is the direct sum of the eigenspaces $E(\lambda)$, $\lambda \in \Lambda$. To prove this statement without the restrictions on the multiplicities, we need

THEOREM 4.3.5. *In every eigenspace* $E(\lambda)$ *of a locally definite problem (4.1.1), there is a* φ*-orthogonal basis consisting of decomposable tensors.*

Proof. Let d be a positive integer. We prove by induction on d that the statement of the theorem holds whenever $\dim E(\lambda) \leq d$. This is trivially true if $d = 1$. Now let us assume that the above statement holds for a given d, and let $\lambda \in \Lambda$ be an eigenvalue of (4.1.1) of multiplicity $d + 1$. Without loss of generality, we shall assume that $E(\lambda) = H$ because we can go over to the projected problem (2.2.7) with $\widetilde{H}_r = E_r(\lambda)$. It should be noted that the sesquilinear forms φ_s belonging to problem (2.2.7) are the restrictions of the forms φ_s to $E(\lambda)$.

By Lemma 4.3.1, we then have

$$\lambda_s \varphi_t = \lambda_t \varphi_s \quad \text{for every} \quad s,t = 0,\ldots,k \ . \tag{4.3.3}$$

We choose ℓ such that H_ℓ has dimension greater than one and t such that λ_t is nonzero. We further choose a Hermitian operator A on H_ℓ with kernel different from $\{0\}$ and H_ℓ. Then we consider the eigenvalue problem

$$\sum_{s=0}^{k} \mu_s \tilde{A}_{rs} x_r = 0 \quad , \quad 0 \neq x_r \in H_r, \quad r = 1,\dots,k \quad , \tag{4.3.4}$$

where

$$\tilde{A}_{rs} = \begin{cases} A_{rs} & \text{if } (r,s) \neq (\ell,t) \ , \\ A_{rs} + A & \text{if } (r,s) = (\ell,t) \ . \end{cases}$$

Since λ_t is nonzero, (4.3.2) and (4.3.3) show that $\varphi_t(x,x) \neq 0$ for all nonzero decomposable tensors x. Since $\tilde{\varphi}_t = \varphi_t$, it follows that problem (4.3.4) is locally definite. Here and in the sequel objects indicated by tilde belong to (4.3.4). We see that $\lambda \in \tilde{\Lambda}$ but the multiplicity of λ with respect to (4.3.4) is smaller than $\dim H = d + 1$. By Lemma 4.3.3, the sum of the multiplicities of the eigenvalues $\mu \in \tilde{\Lambda}$ is equal to $d + 1$. Hence every eigenvalue $\mu \in \tilde{\Lambda}$ has multiplicity smaller than $d + 1$. By the induction hypothesis, there is a $\tilde{\varphi}$-orthogonal basis of decomposable tensors in every eigenspace $\tilde{E}(\mu)$, $\mu \in \tilde{\Lambda}$. If we combine these bases to a system x^m, $m = 1,\dots,\dim H$, then, by Lemma 4.3.2, this system is $\tilde{\varphi}$-orthogonal. By Lemma 4.3.4, the system x^m, $m = 1,\dots,\dim H$, is linearly independent and thus forms a basis of H. Since $\tilde{\varphi}_t = \varphi_t$, it follows from (4.3.3) that the basis x^m is φ-orthogonal. Hence this basis has all desired properties. This completes the induction and the proof of the lemma.□

The above theorem has some interesting consequences. For fixed $\lambda \in \Lambda$, choose a φ-orthogonal basis x^m of $E(\lambda)$ consisting of decomposable tensors. Since λ has positive signum, $\varphi(x^m,x^m)$ is a positive multiple of λ. By normalization of x^m, we can assume that $\varphi(x^m,x^m) = \lambda$ for each m. Every eigenvector $x \in E(\lambda)$ can be written as

$$x = \sum_m \alpha_m x^m \ ,$$

which, by φ-orthogonality of x^m, gives

$$\varphi(x,x) = \left(\sum_m |\alpha_m|^2 \right) \lambda \quad .$$

Hence

$$\varphi(x,x)\,\lambda > 0 \quad \text{for all} \quad 0 \neq x \in E(\lambda) \quad . \tag{4.3.5}$$

In particular, this shows that

$$\psi = \sum_{s=0}^{k} \lambda_s \varphi_s$$

is an inner product in $E(\lambda)$. It follows from Lemma 4.3.1 that $0 = \psi(x,y) = \varphi(x,y)\lambda$ implies $\varphi(x,y) = 0$ whenever $x,y \in E(\lambda)$. Hence φ - orthogonality and ψ - orthogonality are equivalent within $E(\lambda)$. The last remarks motivate the notion "locally definite".

Combining φ - orthogonal bases of every eigenspace $E(\lambda)$, $\lambda \in \Lambda$, we obtain a φ - orthogonal system x^m, $m = 1,\ldots,\dim H$; see Lemmas 4.3.2, 4.3.3. By Lemma 4.3.4, this system is a basis of H . Thus we have proved

THEOREM 4.3.6. *Let the eigenvalue problem (4.1.1) be locally definite. Then* H *is the direct sum of the eigenspaces* $E(\lambda)$, $\lambda \in \Lambda$ *, and there is a* φ *- orthogonal basis* x^m, $m = 1,\ldots,\dim H$ *, of* H *consisting of decomposable eigenvectors.*

We note that if we expand a tensor $x \in H$ with respect to a φ - orthogonal basis of eigenvectors

$$x = \sum_m \alpha_m x^m \tag{4.3.6}$$

then the coefficients α_m are determined by

$$\varphi(x,x^m) = \alpha_m \varphi(x^m,x^m) \quad . \tag{4.3.7}$$

Moreover, we have

$$\varphi(x,x) = \sum_m |\alpha_m|^2 \varphi(x^m,x^m) \quad . \tag{4.3.8}$$

If we normalize x^m such that

$$\varphi(x^m,x^m) = \lambda^m \in \Lambda \quad \text{for each} \quad m \tag{4.3.9}$$

then

$$\varphi(x,x) = \sum_m |\alpha_m|^2 \lambda^m . \tag{4.3.10}$$

For later use, we note

THEOREM 4.3.7. *Let the eigenvalue problem (4.1.1) be locally definite, and let* $\mu = (\mu_0,\ldots,\mu_k)$ *be a nonzero* (k+1)*-tuple of real numbers. Take a* φ*-orthogonal basis of every eigenspace* $E(\lambda)$ *for which* $\lambda \in S^k$ *and* $\mu\lambda$ *is positive, and combine these bases to a system* x^m *. Then*

$$\mu\,\varphi(x,x) = \sum_m \frac{|\mu\,\varphi(x,x^m)|^2}{\mu\,\varphi(x^m,x^m)} \qquad \text{for all} \quad x \in H .$$

Proof. It is possible that (4.1.1) has eigenvalues $\lambda \in \Lambda$ such that $\mu\lambda$ vanishes. Choose a φ-orthogonal basis in each eigenspace associated with such an eigenvalue, and combine these bases to a system y^n . By Lemma 4.3.2 and Theorem 4.3.6, the systems x^m and y^n together form a φ-orthogonal basis of H . Hence a given $x \in H$ can be expanded as

$$x = \sum_m \alpha_m x^m + \sum_n \beta_n y^n .$$

Since $\varphi(x^m,x^m)$ is a nonzero multiple of an eigenvalue λ satisfying $\mu\lambda > 0$, $\mu\,\varphi(x^m,x^m)$ is nonzero. Hence (4.3.7) gives $\alpha_m = \mu\,\varphi(x,x^m)/\,\mu\,\varphi(x^m,x^m)$, and Lemma 4.3.1 shows that $\mu\,\varphi(y^n,y^n)$ vanishes. Now multiplying

$$\varphi(x,x) = \sum_m |\alpha_m|^2 \varphi(x^m,x^m) + \sum_n |\beta_n|^2 \varphi(y^n,y^n)$$

by μ , we obtain the statement of our theorem. □

4.4 Positive definite form determinants

We draw some conclusions from the results of Section 4.3 concerning a form determinant

$$\psi = \det_{1\le r,s\le k} \psi_{rs} ,$$

where ψ_{rs} is a Hermitian sesquilinear form on a finite dimensional space H_r for

$r,s = 1,\dots,k$. If ψ is positive definite then, by (4.2.3),

$$\det_{1\le r,s\le k} \psi_{rs}(x_r,x_r) > 0 \quad \text{for all} \quad 0 \neq x_r \in H_r \quad . \tag{4.4.1}$$

It is remarkable that also the converse statement is true.

THEOREM 4.4.1. *Assume that (4.4.1) holds. Then*

(i) ψ *is positive definite, and*

(ii) there is a basis of $H = H_1 \otimes \dots \otimes H_k$ *, orthonormal with respect to* ψ *, consisting of decomposable tensors.*

Proof. We choose inner products $<\,,\,>_r$ and Hermitian operators A_{rs} on H_r such that $\psi_{rs} = <A_{rs}\,\cdot,\cdot>_r$, $r,s = 1,\dots,k$. We further set $A_{ro} := 0$ for each r . With these operators we consider the eigenvalue problem (4.1.1) which is locally definite because $\varphi_0 = \psi$ and $\varphi_s = 0$, $s = 1,\dots,k$. We have $H = E(\lambda)$, where $\lambda = (1,0,\dots,0)$ is the only eigenvalue in S^k which has positive signum. Hence (i) follows from (4.3.5) and (ii) from Theorem 4.3.5 because φ-orthogonality and ψ-orthogonality are equivalent.□

4.5 The expansion theorem for definite problems

We consider the eigenvalue problem (4.1.1). If this problem is definite according to Section 1.5 then there is a (k+1)-tuple of real numbers $\mu = (\mu_0,\dots,\mu_k)$ such that

$$\mu\,\varphi(x,x) > 0 \quad \text{for all nonzero decomposable } x \quad . \tag{4.5.1}$$

The sesquilinear form

$$\psi = \sum_{s=0}^{k} \mu_s \varphi_s \tag{4.5.2}$$

is the form determinant of the $k+1$ by $k+1$ array

$$\begin{bmatrix} \mu_o & \dots & \mu_k \\ <A_{1o}\,\cdot,\cdot>_1 & \dots & <A_{1k}\,\cdot,\cdot>_1 \\ \vdots & & \vdots \\ <A_{ko}\,\cdot,\cdot>_k & \dots & <A_{kk}\,\cdot,\cdot>_k \end{bmatrix} \tag{4.5.3}$$

of Hermitian sesquilinear forms. Thereby we identify $\mathbb{C} \otimes H_1 \otimes \ldots \otimes H_k$ with $H_1 \otimes \ldots \otimes H_k$ and the number μ_s with the form $(x,y) \mapsto \mu_s\, x\bar{y}$. Hence (4.5.1) and Theorem 4.4.1 show that ψ is an inner product in H . We can now formulate a variant of the expansion Theorems 4.3.6, 4.3.7.

THEOREM 4.5.1. *Let the eigenvalue problem (4.1.1) be definite with respect to* μ *, and let* ψ *be the inner product defined by (4.5.2). Then there is a basis* x^m, $m = 1,\ldots,\dim H$ *, of* H *consisting of decomposable eigenvectors which is orthonormal with respect to* ψ *. For every* $x \in H$ *, there holds the Fourier expansion*

$$x = \sum_m \psi(x, x^m)\, x^m ,$$

and Parseval's identity

$$\psi(x,x) = \sum_m |\psi(x,x^m)|^2 .$$

We recall that (4.1.1) is locally definite if and only if (4.3.2) holds. There is a similar criterion for (4.1.1) to be definite.

THEOREM 4.5.2. *The eigenvalue problem (4.1.1) is definite if and only if*

$$\varphi(x,x) \neq 0 \quad \textit{for all} \quad 0 \neq x \in H . \tag{4.5.4}$$

Proof. If (4.1.1) is definite with respect to μ then, as remarked above, $\mu\varphi(x,x)$ is positive for all nonzero tensors x . This proves (4.5.4).

Now assume that (4.5.4) holds. Then (4.1.1) is locally definite. By Theorem 4.3.6, there is a φ- orthogonal basis x^m of H consisting of decomoposable eigenvectors. We normalize this basis such that (4.3.9) holds. Then, by (4.3.10), the set

$$\{\varphi(x,x) \mid x = \sum_m \alpha_m x^m , \ \sum_m |\alpha_m|^2 = 1\} \tag{4.5.5}$$

is the convex hull of the finite set Λ . Hence the set (4.5.5) is compact and convex and, by assumption (4.5.4), it does not contain the zero vector. As is well known, such a set lies in an open halfspace of $\mathbb{R}^{k+1}$ determined by a hyperplane through zero. Hence there is a tuple $\mu = (\mu_0,\ldots,\mu_k) \in \mathbb{R}^{k+1}$ such that $\mu\varphi(x,x)$ is positive for all $0 \neq x \in H$ which proves definiteness.□

4.6 Notes for Chapter 4

Atkinson (1972) studied the eigenvalue problem (4.1.1) via certain operator determinants Δ_s , $s = 0,\dots,k$, defined on the tensor product H . These operators Δ_s are related to the forms φ_s by

$$<\Delta_s x,y> = \varphi_s(x,y) \quad , \quad x,y \in H \quad ,$$

where $< , >$ denotes the tensorial product of the inner products $< , >_r$. It should be noted that we did not use these operators Δ_s and the inner product $< , >$ in H . Atkinson reduced the locally definite problem (4.1.1) to the simultaneous eigenvalue problem

$$\Gamma_s x = \lambda_s x \quad , \quad 0 \neq x \in H \ , \ s = 0,\dots,k \quad ,$$

where

$$\Gamma_s = \Delta^{-1}\Delta_s$$

and Δ is a suitable linear combination of the operators Δ_s, $s = 0,\dots,k$. It can be shown that the operators $\Gamma_0,\dots,\Gamma_k$ are pairwise commutative; see [Atkinson (1972), Theorem 6.7.2]. Hence the system $\Gamma_0,\dots,\Gamma_k$ has at least one simultaneous eigenvalue. This result replaces our Theorem 1.4.1 in Atkinson's proof of Theorem 4.3.6; see [Atkinson (1972), Chapter 10].

Another approach to Theorem 4.3.6 was given by Binding and Browne (1977).

The problem of multiple eigenvalues arising in the proof of Theorem 4.3.6 was solved by Atkinson using a perturbation theorem. This method is slightly different from ours.

Concerning Theorem 4.4.1(i) we refer to Theorem 7.8.2 of Atkinson (1972); compare also Binding (1984d) and Volkmer (1986). The expansion Theorem 4.5.1 in the definite case is Theorem 7.9.1 of Atkinson (1972).

CHAPTER 5

MULTIPARAMETER EXPANSION THEOREMS FOR BOUNDED OPERATORS

5.1 Introduction

We suppose given k nonzero separable Hilbert spaces $(H_r, <\ ,\ >_r)$, $r = 1,\ldots,k$. For each r, let A_r and A_{rs}, $s = 1,\ldots,k$, be compact Hermitian operators on H_r. We set $A_{ro} := A_r - I_r$, $r = 1,\ldots,k$, where I_r denotes the identity operator on H_r.

We shall study multiparameter expansion theorems for the eigenvalue problem of Chapter 2

$$\sum_{s=0}^{k} \lambda_s A_{rs} x_r = 0,\ \lambda_0 = 1,\ 0 \neq x_r \in H_r,\ r = 1,\ldots,k\ . \tag{5.1.1}$$

After the preparatory Section 5.2, we define eigenvectors and eigenspaces of problem (5.1.1) in Section 5.3. These definitions are completely analogous to those in the finite dimensional case.

The main idea of this chapter is to approximate problem (5.1.1) by a sequence of projected problems (2.2.7) and then to carry over the expansion theorem of Chapter 4 to the limiting problem (5.1.1). This approximation method is explained in Section 5.4. It leads to an expansion theorem for (5.1.1) provided we are able to verify that certain limiting processes can be interchanged. This is a rather delicate task which will be treated for left definite and for right definite eigenvalue problems.

The left definite case is studied in Sections 5.5, 5.6, 5.7. The results depend on simple properties of Minkowski matrices which are given in Section 5.5. Then we carry out the approximation procedure leading to an expansion theorem for the sesquilinear form φ_0; see Section 5.6. In the next section we show that a left definite problem is closely related to the eigenvalue problem for a certain compact Hermitean operator acting on the tensor product of the spaces H_r.

Expansion theorems for right definite eigenvalue problems are studied in Section 5.8. In Section 5.9 we investigate a multiparameter eigenvalue problem for integral operators as a special case of our problem (5.1.1). In particular, we refine the statements of our general expansion theorems in this case. References to the literature are collected in Section 5.10.

5.2 A result on form determinants

Let H_r, $r = 1,\dots,k$, be complex linear spaces of arbitrary dimension, and let

$$H_1 \times \dots \times H_k \ni (x_1,\dots,x_k) \mapsto x_1 \otimes \dots \otimes x_k \in H := H_1 \otimes \dots \otimes H_k \qquad (5.2.1)$$

be the canonical map associated with their tensor product H . We recall that H is the linear span of the range of the map (5.2.1) i. e. every tensor is a finite sum of decomposable tensors. In particular, every tensor is contained in a tensor product $F_1 \otimes \dots \otimes F_k$ of finite dimensional subspaces F_r of H_r .

If ψ_r, $r = 1,\dots,k$, are sesquilinear forms on H_r then their *tensorial product* $\psi = \psi_1 \otimes \dots \otimes \psi_k$ is the sesquilinear form on H which is uniquely determined by property (4.2.1). If ψ_r is Hermitian for each r then ψ is Hermitian, too.

Now let ψ_{rs}, $r,s = 1,\dots,k$, be a square array of sesquilinear forms on H_r . Then their *form determinant* ψ is defined by (4.2.2). We remark that the following generalization of Theorem 4.4.1(i) holds.

THEOREM 5.2.1. *Let* ψ *be the form determinant of the Hermitian sesquilinear forms* ψ_{rs}, $r,s = 1,\dots,k$. *Then, if* $\psi(x,x)$ *is positive for all nonzero decomposable tensors,* ψ *is positive definite.*

Proof. Let x be a nonzero tensor. Then there are finite dimensional subspaces F_r of H_r such that $x \in F := F_1 \otimes \dots \otimes F_k$. The restriction of ψ to F is the form determinant of the restrictions of ψ_{rs} to F_r . Hence, by Theorem 4.4.1(i), the restriction of ψ to F is positive definite, and thus $\psi(x,x)$ is positive.□

5.3 Eigenvectors and eigenspaces

We consider the eigenvalue problem (5.1.1). If $\lambda = (1,\lambda_1,\dots,\lambda_k)$ is an eigenvalue then we define the *eigenspace* $E(\lambda)$ to be the tensor product

$$E(\lambda) := E_1(\lambda) \otimes \dots \otimes E_k(\lambda) \subset H := H_1 \otimes \dots \otimes H_k \quad ,$$

where $E_r(\lambda)$ is the eigenspace of the compact operator

$$A_r + \sum_{s=1}^{k} \lambda_s A_{rs}$$

associated with the eigenvalue 1 . The nonzero tensors in $E(\lambda)$ are the *eigenvectors* belonging to λ . The dimension of the eigenspace $E(\lambda)$ is finite. If λ has real components then we see as in Section 4.3 that the dimension of $E(\lambda)$ is equal to the multiplicity of λ .

For $x,y \in H$, we set

$$\varphi(x,y) := (\varphi_0(x,y),\ldots,\varphi_k(x,y)) \quad ,$$

where $(-1)^s\varphi_s$ is the form determinant of the array (4.3.1). If $x = x_1 \otimes \ldots \otimes x_k$ is a decomposable eigenvector belonging to the eigenvalue λ and $y = y_1 \otimes \ldots \otimes y_k$ is an arbitrary decomposable tensor then we see as in the proof of Lemma 4.3.1 that $\varphi(x,y)$ is a multiple of λ . This remains true for all $x \in E(\lambda)$ and $y \in H$. Hence, noting that $\lambda_0 = 1$, we obtain

$$\varphi_s(x,y) = \lambda_s\varphi_0(x,y) \ , \quad x \in E(\lambda) \ , \ y \in H \ , \ s = 1,\ldots,k \quad . \tag{5.3.1}$$

Some elementary consequences of local definiteness are collected in

LEMMA 5.3.1. *Let the eigenvalue problem (5.1.1) be locally definite. Then the following statements hold.*

(i) Eigenspaces belonging to different eigenvalues are φ*-orthogonal i. e.*

$$\varphi(x,y) = 0 \quad \textit{whenever} \quad x \in E(\lambda),\ y \in E(\mu),\ \lambda \neq \mu \quad .$$

(ii) In every eigenspace there is a φ*-orthogonal basis consisting of decomposable tensors.*

(iii) In every eigenspace there is an inner product such that orthogonality with respect to this inner product is equivalent to φ*-orthogonality within that eigenspace.*

(iv) The eigenspaces form a direct sum.

PROOF. (i) Since (5.1.1) is locally definite, the eigenvalues of (5.1.1) have real components. Hence, by (5.3.1), $\varphi(x,y)$ is a multiple of both λ and μ . Since

$\lambda_0 = \mu_0 = 1$ and $\lambda \neq \mu$, the eigenvalues λ, μ are linearly independent. Thus $\varphi(x,y)$ is zero.

(ii) follows from Theorem 4.3.5 applied to the projected problem (2.2.7) with $\tilde{H}_r = E_r(\lambda)$.

(iii) is an easy consequence of (ii) as shown after the proof of Theorem 4.3.5.

(iv) follows from (i),(ii) and Lemma 4.3.4 which remains valid for the eigenvalue problem (5.1.1).□

For the formulation of the expansion theorems in the next sections, let us fix some notations. If the eigenvalue problem (5.1.1) is locally definite then, by Theorem 2.2.3, there is at most one eigenvalue of problem (5.1.1) of a given signed index $j = (i,\sigma)$. Let J denote the set of all signed indices j such that the eigenvalue λ^j of (5.1.1) of signed index j exists. An *eigensystem* of (5.1.1) is defined to be a φ-orthogonal system $x^j \in E(\lambda^j)$, $j \in J$, of eigenvectors. Since the dimension of $E(\lambda^j)$ is equal to the multiplicity of λ^j, Lemma 5.3.1(i),(ii) shows that we can always find an eigensystem consisting of decomposable tensors. We remark that decomposability is not part of the definition of eigensystems. Lemma 5.3.1(iii),(iv) shows that eigensystems are linearly independent and contain bases of every eigenspace of (5.1.1).

5.4 Approximation of eigenvalue problems

In this section we assume that the eigenvalue problem (5.1.1) is locally definite in the strong sense. We approximate this problem by a sequence of projected problems

$$\sum_{s=0}^{k} \lambda_s P_r^n A_{rs} x_r = 0, \quad \lambda_0 = 1, \; 0 \neq x_r \in H_r^n, \; r = 1,\dots,k, \tag{5.4.1}$$

where n runs through the set of positive integers, and, for each r and n, P_r^n is an orthoprojector on H_r with range H_r^n. We assume that the following two conditions are satisfied.

1) For each r and n, the space H_r^n is finite dimensional.
2) For each r, the sequence $H_r^1, H_r^2, \dots$ is increasing. Their union is dense in H_r, or, equivalently, $P_r^n x_r$ converges to x_r as $n \to \infty$ for all $x_r \in H_r$.

Such spaces H_r^n exist because H_r is separable for each r. We remark that the forms $\langle P_r^n A_{rs}\cdot,\cdot\rangle_r$ coincide with $\langle A_{rs}\cdot,\cdot\rangle_r$ on H_r^n. Hence the eigenvalue problem (5.4.1) is also locally definite in the strong sense for each n. Similarly, the forms φ_s belonging to (5.4.1) coincide with those belonging to (5.1.1) on $H^n := H_1^n \otimes \ldots \otimes H_k^n$.

We recall that the eigenvalues of (5.1.1) are given by λ^j, $j \in J$, where λ^j is the eigenvalue of signed index j. Analogously, for each n, the eigenvalues of (5.4.1) are written as λ^{jn}, $j \in J^n$. It follows from Theorem 2.5.3 and conditions 1),2) that the sets J^n are finite subsets of J and increase with n:

$$J^1 \subset J^2 \subset J^3 \subset \ldots \quad \subset J \ . \tag{5.4.2}$$

Using the notation

$$|j| := \max\{i_1,\ldots,i_k\} \quad \text{for} \quad j = (i,\sigma),\ i = (i_1,\ldots,i_k) \ ,$$

we have the following

LEMMA 5.4.1. *Let* Ω *be a bounded subset of* $\mathbb{R}^{k+1}$. *Then there is a positive integer* m *such that* λ^{jn} *is not contained in* Ω *whenever* $j \in J^n$ *satisfies* $|j| \geq m$.

Proof. Assume the contrary. Then there are positive integers

$$n(1) < n(2) < n(3) < \ldots$$

and a sequence

$$j(m) = (i(m),\sigma(m)) \in J^{n(m)} \ , \quad m = 1,2,\ldots$$

such that, for at least one $\ell \in \{1,\ldots,k\}$,

$$i_\ell(m) \geq m \quad \text{for each} \quad m \ ,$$

and

$$\lambda(m) := \lambda^{j(m)n(m)} \in \Omega \quad \text{for each} \quad m \ .$$

Since Ω is bounded, we can additionally assume that $\lambda(m)$ converges in $\mathbb{R}^{k+1}$ as $m \to \infty$. Then, by condition 2) and Lemma 2.8.3, the operators

$$P_\ell^{n(m)} A_\ell P_\ell^{n(m)} + \sum_{s=1}^{k} \lambda_s(m) P_\ell^{n(m)} A_{\ell s} P_\ell^{n(m)} \tag{5.4.3}$$

converge to a compact Hermitian operator B on H_ℓ as $m \to \infty$ with respect to the operator norm. The $i_\ell(m)$th greatest eigenvalue of (5.4.3), counted according to multiplicity, is equal to 1. Since $i_\ell(m) \geq m$, it follows from Lemma 2.8.1 that B has infinitely many eigenvalues not smaller that 1. This contradicts compactness of B. □

In the sequel, we shall assume that, in addition to 1) and 2), the following third condition is satisfied.

3) For all n and $j \in J$, the inequality $|j| \leq n$ implies $j \in J^n$, $\lambda^j = \lambda^{jn}$ and $E(\lambda^j) \subset H^n$.

There are spaces H_r^n which satisfy all conditions 1), 2), 3). In fact, by Lemma 2.2.2, we can find finite dimensional subspaces H_r^n, $r = 1,\dots,k$, n fixed, such that $|j| \leq n$ implies $j \in J^n$ and $\lambda^j = \lambda^{jn}$. Obviously, we can successively choose these spaces for $n = 1,2,\dots$ in such a way that also condition 2) is satisfied and that the eigenspace $E(\lambda^j)$ of (5.1.1) is contained in H^n if $|j| \leq n$.

Now let x^j, $j \in J$, be a fixed eigensystem of (5.1.1). Then x^j is an eigenvector of (5.1.1) belonging to the eigenvalue λ^j. It follows from condition 3) that x^j is also an eigenvector of (5.4.1) belonging to the eigenvalue $\lambda^{jn} = \lambda^j$ if $|j| \leq n$. Hence, by Lemma 5.3.1(i),(iii), we can complete the φ-orthogonal system x^j, $j \in J$ and $|j| \leq n$, to an eigensystem x^{jn}, $j \in J^n$, of (5.4.1). Thus

$$j \in J^n \text{ and } x^{jn} = x^j \text{ whenever } j \in J \text{ and } |j| \leq n . \tag{5.4.4}$$

By condition 1) and Theorem 4.3.7 with $\mu = (1,0,\dots,0)$, we have

$$\varphi_0(x,x) = \sum_{j \in J^n} \frac{|\varphi_0(x,x^{jn})|^2}{\varphi_0(x^{jn},x^{jn})} \quad \text{for all } x \in H^n . \tag{5.4.5}$$

We now prove that, under suitable assumptions, we can go to the limit $n \to \infty$ in formula (5.4.5). We remark that, for any given $x \in H$, we can find an approximation of problem (5.1.1) which satisfies conditions 1), 2), 3) such that x lies in H^1, and therefore lies in H^n for each n.

THEOREM 5.4.2. *Let* $x \in H^1$, *and let the series*

$$\sum_{j \in J^n} \frac{|\varphi_0(x, x^{jn})|^2}{\varphi_0(x^{jn}, x^{jn})} \quad , \; n = 1,2,\ldots \tag{5.4.6}$$

be uniformly and absolutely convergent i.e., for every positive ε, *there is a positive integer* m *such that*

$$\sum_{\substack{j \in J^n \\ |j| \geq m}} \frac{|\varphi_0(x, x^{jn})|^2}{|\varphi_0(x^{jn}, x^{jn})|} < \varepsilon \quad \textit{for each } n \; . \tag{5.4.7}$$

Then $\varphi_0(x,x)$ *admits the representation*

$$\varphi_0(x,x) = \sum_{j \in J} \frac{|\varphi_0(x, x^j)|^2}{\varphi_0(x^j, x^j)} \; , \tag{5.4.8}$$

where the series is absolutely convergent.

Proof. Let ε be positive, and choose m such that (5.4.7) holds. Then (5.4.4) shows that

$$\sum_{\substack{j \in J \\ m \leq |j| \leq n}} \frac{|\varphi_0(x, x^j)|^2}{|\varphi_0(x^j, x^j)|} < \varepsilon \quad \text{for each } n$$

which proves absolute convergence of the series (5.4.8). It follows from (5.4.2), (5.4.4),(5.4.5) that

$$\varphi_0(x,x) - \sum_{\substack{j \in J \\ |j| \leq n}} \frac{|\varphi_0(x, x^j)|^2}{\varphi_0(x^j, x^j)} = \sum_{\substack{j \in J^n \\ |j| > n}} \frac{|\varphi_0(x, x^{jn})|^2}{\varphi_0(x^{jn}, x^{jn})} \; .$$

By (5.4.7), the absolute value of this number is smaller than ε provided that n is greater than m. This proves (5.4.8).□

This is our preliminary expansion theorem for the form φ_0. We shall give sufficient conditions for the series (5.4.6) to be uniformly and absolutely convergent in Sections 5.6 and 5.8.

5.5 Minkowski matrices

For the proof of the expansion theorem in the left definite case, we need some properties of real matrices. We first show

LEMMA 5.5.1. *Let* A *be a real* k *by* k *matrix such that all offdiagonal elements are nonpositive. Then the following three statements are equivalent.*

(i) There is a column vector $x \in \mathbb{R}^k$ *such that all components of* x *and* Ax *are positive.*

(ii) All principal minors of A *are positive.*

(iii) The matrix A *is invertible and all elements of its inverse matrix* A^{-1} *are nonnegative.*

Proof. (i) implies (ii): Let us denote the matrix elements of A by a_{rs}, $r,s = 1,\dots,k$. Then there are positive numbers $x_1,\dots,x_k$ such that

$$a_{rr}\,x_r > -\sum_{s\neq r} a_{rs}\,x_s = \sum_{s\neq r} |a_{rs}\,x_s| \quad , \; r = 1,\dots,k \; .$$

Hence the matrix $a_{rs}\,x_s$, $r,s = 1,\dots,k$, is diagonally dominant and therefore invertible. It follows that the matrix A is invertible, too. Statement (i) remains true if we replace A by $tE + (1-t)A$, $0 \leq t \leq 1$, E denoting the unit matrix. Hence $\det(tE + (1-t)A)$ is nonzero for all $0 \leq t \leq 1$. Consequently, $\det A$ is positive. Since (i) remains true if we replace A by one of its principal submatrices, we obtain (ii).

(ii) implies (iii): We prove this implication by induction on k. The case $k = 1$ is trivial. Now let us assume that the implication holds with $k-1$ in place of k, and let A be a real k by k matrix with nonpositive offdiagonal elements which satisfies (ii). Let b_{rs} denote the cofactors of A i. e. $(-1)^{r+s}b_{rs}$ is the determinant of the matrix A with r^{th} row and s^{th} column deleted. It is well known that $b_{sr}/\det A$ is the element of A^{-1} in the rth row and s^{th}

column. By assumption (ii), this element is positive if $r = s$. It remains to show that b_{rs} is nonnegative if $r \neq s$. To simplify the notation, we assume that $r = k$. By expanding the determinant $(-1)^{k+s} b_{ks}$, $s \neq k$, according to the last column, we obtain

$$(-1)^{k+s} b_{ks} = \sum_{r=1}^{k-1} a_{rk} (-1)^{r+k-1} (-1)^{r+s} c_{rs} \quad ,$$

where c_{rs} denotes the cofactors of the matrix A with k^{th} row and column deleted. By the induction hypothesis, c_{rs} is nonnegative for every $r,s=1,\dots,k-1$. Since a_{rk} is nonpositive for every $r = 1,\dots,k-1$, it follows that b_{ks} is nonnegative.

(iii) implies (i): Let x be the vector sum of the columns of A^{-1}. Since all elements of A^{-1} are nonnegative, the components of x are nonnegative, too. In fact, these components are positive because A^{-1} is invertible. Since $A A^{-1} = E$, every component of Ax is equal to 1. □

A real k by k matrix A with nonpositive offdiagonal elements which satisfies the equivalent conditions (i),(ii),(iii) of the preceding lemma is called a *non-singular M-matrix* by [Berman and Plemmons (1979), Chapter 6, Theorem (2.3)].

LEMMA 5.5.2. *Let* A *be a real* k *by* k *matrix with nonpositive offdiagonal elements. Then, if there is a column vector* x *with positive components satisfying* $Ax = 0$, *the convex hull of the rows of* A *contains the zero vector.*

Proof. For all positive ε, $A + \varepsilon E$ is a nonsingular M-matrix because condition (i) of Lemma 5.5.1 is satisfied. Hence the matrix $(A + \varepsilon E)^{-1}$ has nonnegative elements, and the diagonal elements are positive. Let α_n be the maximal row sum of $(A + 1/nE)^{-1}$, and let $B_n := 1/\alpha_n (A + 1/nE)^{-1}$. Then

$$B_n(A + 1/nE) = 1/\alpha_n E \quad \text{for each } n \ .$$

The sequence B_n has a subsequence converging to a matrix B. We see that BA is a multiple of E. Since A is singular, it follows that $BA = 0$. Hence the zero vector is a convex combination of the rows of A because B has nonnegative elements and the maximal row sum of B is equal to 1. □

LEMMA 5.5.3. *Let* F *be a real linear space endowed with an inner product* $\langle\, ,\, \rangle$. *Further, let* $a_1,\dots,a_k$ *be vectors generating* F *such that the zero vector is a linear combination of* $a_1,\dots,a_k$ *with positive coefficients, and let* $b_1,\dots,b_k$ *be vectors in* F *satisfying*

$$\langle a_r,b_s\rangle \le 0 \quad \textit{whenever} \quad r \neq s \ .$$

Then the convex hull of the set $\{b_1,\dots,b_k\}$ *contains the zero vector.*

Proof. The transposed matrix of $\langle a_r,b_s\rangle$, $r,s = 1,\dots,k$, satisfies the assumptions of Lemma 5.5.2. Hence there are nonnegative numbers $y_1,\dots,y_k$ whose sum is equal to 1 such that

$$\sum_{s=1}^{k} y_s \langle a_r,b_s\rangle = 0 \quad \text{for every} \quad r = 1,\dots,k \ .$$

Since $a_1,\dots,a_k$ generate F, we obtain $\Sigma\, y_s b_s = 0$, which proves the lemma. □

5.6 The expansion theorem for left definite problems

We consider the eigenvalue problem (5.1.1) which is assumed to be left definite with respect to $\mu_1,\dots,\mu_k$; see Section 2.7. Then, by Lemma 2.7.2, our problem (5.1.1) is definite with respect to $\mu = (0,\mu_1,\dots,\mu_k)$. In particular, the sesquilinear form

$$\psi = \sum_{s=1}^{k} \mu_s \varphi_s \tag{5.6.1}$$

has the property that $\psi(x,x)$ is positive for all nonzero decomposable tensors. Since ψ is the form determinant of the array (4.5.3), it follows from Theorem 5.2.1 that ψ is an inner product in H.

The left definite eigenvalue problem (5.1.1) is closely related to a family of right definite problems in the following way. We replace one of the k Hilbert spaces $H_1,\dots,H_k$, say H_ℓ, by the complex field and the ℓth equation of (5.1.1) by the scalar equation $-\tau + \mu\lambda = 0$, where τ is a real constant. Then, by (2.7.2), we obtain a right definite eigenvalue problem depending on ℓ and τ. By Theorem 2.7.1, this right definite problem has the eigenvalue of signed index $((1,\dots,1),1)$ which we denote by $\lambda(\ell,\tau)$. The eigenvalue $\lambda = \lambda(\ell,\tau)$ is uniquely determined by its properties $\mu\lambda = \tau$ and

$$0 = \max\{w_r(u_r)\lambda \mid u_r \in U_r\} \quad \text{if} \quad r \neq \ell \ . \tag{5.6.2}$$

We recall that $w_r(u_r)$ denotes the rth row of the matrix $W(u)$ defined by (1.1.2), and U_r is the unit sphere of H_r .

From Theorem 2.9.1 we deduce

LEMMA 5.6.1. *The functions* $\lambda(\ell,\cdot)$, $\ell = 1,\dots,k$, *are continuous.*

Using the results of Section 5.5, we obtain

LEMMA 5.6.2. *Let* $\lambda = (1,\lambda_1,\dots,\lambda_k) \in \mathbb{R}^{k+1}$ *and* $u = (u_1,\dots,u_k) \in U = U_1 \times \dots \times U_k$ *satisfy* $W(u)\lambda = 0$. *Then, setting* $\tau := \mu\lambda$, *we have*

$$\lambda \in \operatorname{co}\{ \lambda(\ell,\tau) \mid \ell = 1,\dots,k\} \ .$$

Proof. Let β_ℓ be the positive determinant (2.7.2) for the given unit vectors $u_1,\dots,u_k$. Then

$$\sum_{\ell=1}^{k} \beta_\ell \, w_\ell(u_\ell) \ \in \ \operatorname{span}\{ \mu , a \} \ ,$$

where $a = (1,0,\dots,0)$. Hence if P denotes the orthoprojector on $\mathbb{R}^{k+1}$ whose kernel is spanned by μ and a then the zero vector is a linear combination of $Pw_1(u_1),\dots,Pw_k(u_k)$ with positive coefficients. Moreover, the system $Pw_r(u_r)$, $r = 1,\dots,k$, generates the range of P .

By $W(u)\lambda = 0$ and (5.6.2), we have

$$w_r(u_r)(\lambda(\ell,\tau) - \lambda) = w_r(u_r)\ \lambda(\ell,\tau) \leq 0 \quad \text{if} \quad r \neq \ell \ .$$

By definition of τ , the vectors $\lambda(\ell,\tau) - \lambda$, $\ell = 1,\dots,k$, are contained in the range of P . Hence we see that

$$Pw_r(u_r)(\lambda(\ell,\tau) - \lambda) \leq 0 \quad \text{if} \quad r \neq \ell \ .$$

We can now apply Lemma 5.5.3 to the systems $Pw_r(u_r)$, $r = 1,\dots,k$, and $\lambda(\ell,\tau) - \lambda$, $\ell = 1,\dots,k$. It follows that the zero vector lies in the convex hull of the system $\lambda(\ell,\tau) - \lambda$, $\ell = 1,\dots,k$. This immediately yields the statement of the lemma.□

We mention that the preceding lemma has a simple geometric meaning. As an example, let us consider the case $k = 3$ and $\mu_1 = 0,\ \mu_2 = 0,\ \mu_3 = 1$. For every fixed $\lambda_3 = \tau$ and $r = 1,2,3$, we have eigencurves in the (λ_1,λ_2)-plane defined by

$$\rho_r((1,\lambda_1,\lambda_2,\tau),1) = 0 \tag{5.6.3}$$

i. e. these curves consist of those pairs (λ_1,λ_2) for which the greatest eigenvalue of the operator

$$A_{ro} + \lambda_1 A_{r1} + \lambda_2 A_{r2} + \tau A_{r3}$$

is equal to 0 . The curves are shown in the following picture.

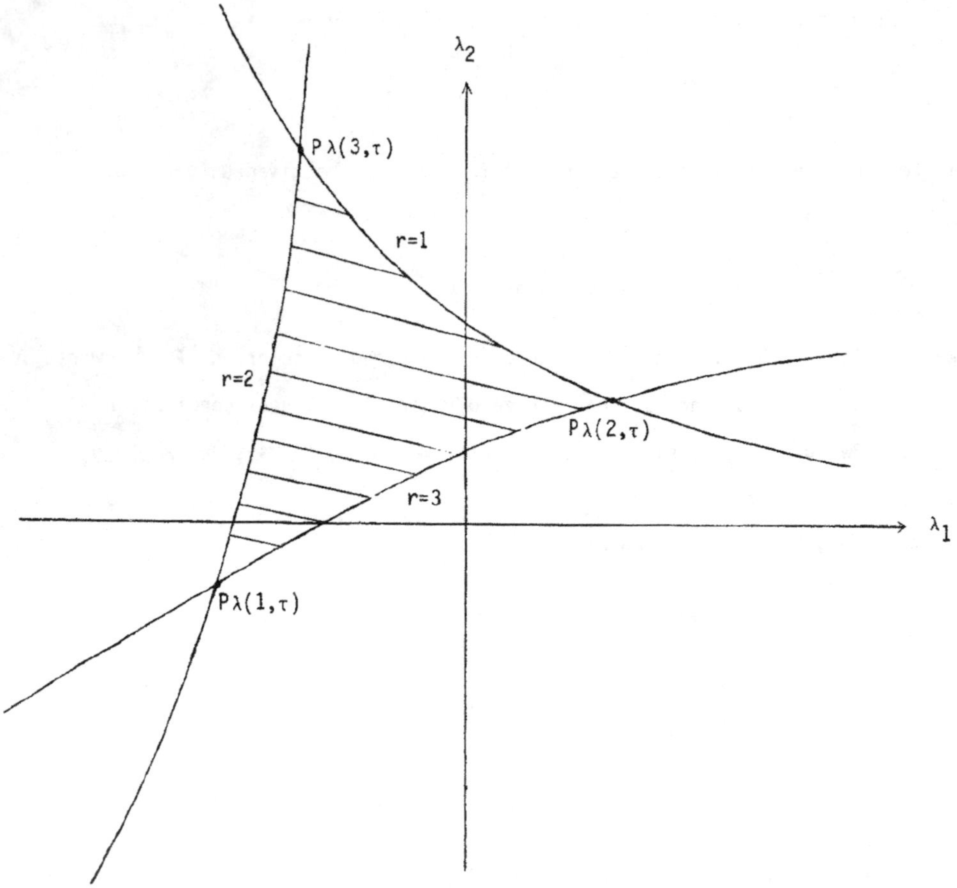

The curves indicated by $r = 1$, $r = 2$, $r = 3$ consist of all solutions (λ_1,λ_2) of equation (5.6.3) for $r = 1,2,3$, respectively. For each $\ell = 1,2,3$,

$$P\lambda(\ell,\tau) = (\lambda_1(\ell,\tau) \;,\; \lambda_2(\ell,\tau))$$

is the point of intersection of the two curves indicated by $r = 1,2,3$, $r \neq \ell$. It is easy to see that the shaded region contains all pairs (λ_1,λ_2) for which $W(u)\lambda = 0$, $\lambda := (1,\lambda_1,\lambda_2,\tau)$, is solvable by suitable $u \in U$. The above lemma states that every such pair lies inside the triangle with corners $P\lambda(\ell,\tau)$, $\ell = 1,2,3$. This statement is weaker but its exact formulation is easier because we do not appeal to curves or surfaces etc.

Lemma 5.6.2 together with Lemma 5.6.1 proves the following important boundedness principle valid for left definite problems.

THEOREM 5.6.3. *For every positive* τ, *the set of all* $\lambda = (1,\lambda_1,\ldots,\lambda_k) \in \mathbb{R}^{k+1}$ *which satisfy* $|\mu\lambda| \leq \tau$ *and solve* $W(u)\lambda = 0$ *for suitable* $u \in U$ *is bounded.*

We use this result to prove the following expansion theorem for the sesquilinear form φ_0.

THEOREM 5.6.4. *Let* x^j, $j \in J$, *be an eigensystem of the left definite problem (5.1.1). Then, for all* $x,y \in H$, $\varphi_0(x,y)$ *can be expanded into the absolutely convergent series*

$$\varphi_0(x,y) = \sum_{j\in J} \frac{\varphi_0(x,x^j)\;\varphi_0(x^j,y)}{\varphi_0(x^j,x^j)} \;, \tag{5.6.4}$$

Proof. We shall apply the results of Section 5.4. For a given $x \in H$, we choose an approximation (5.4.1) of the eigenvalue problem (5.1.1) satisfying the conditions 1), 2), 3) of Section 5.4 such that x is contained in H^1. Further, for each n, let x^{jn}, $j \in J^n$, be an eigensystem of (5.4.1) which is related to the given eigensystem x^j, $j \in J$, according to (5.4.4). Since the eigensystem x^{jn}, $j \in J^n$, is φ-orthogonal for each n, it is also orthogonal with respect to the inner product ψ defined by (5.6.1). Hence Bessel's inequality gives

$$\sum_{j\in J^n} \frac{|\psi(x,x^{jn})|^2}{\psi(x^{jn},x^{jn})} \le \psi(x,x) \quad \text{for each } n . \tag{5.6.5}$$

By (5.3.1), we have

$$\psi(y,x^{jn}) = (\mu\lambda^{jn})\,\varphi_0(y,x^{jn}) \quad , y \in H^n .$$

Hence (5.6.5) shows that

$$\sum_{j\in J^n} |\mu\lambda^{jn}| \frac{|\varphi_0(x,x^{jn})|^2}{|\varphi_0(x^{jn},x^{jn})|} \le \psi(x,x) \quad \text{for each } n . \tag{5.6.6}$$

Now let τ be a given positive number. Since every eigenvalue $\lambda = \lambda^{jn}$ solves $W(u)\lambda = 0$ for suitable $u \in U$, it follows from Theorem 5.6.3 that the set of all eigenvalues λ^{jn}, $j \in J^n$, $n = 1,2,\ldots$, which satisfy the inequality $|\mu\lambda^{jn}| \le \tau$, is bounded. Hence, by Lemma 5.4.1, there is a positive integer m such that

$$|\mu\lambda^{jn}| > \tau \quad \text{whenever } j \in J^n, |j| \ge m . \tag{5.6.7}$$

From (5.6.6), (5.6.7) it follows that

$$\sum_{\substack{j\in J^n \\ |j|\ge m}} \frac{|\varphi_0(x,x^{jn})|^2}{|\varphi_0(x^{jn},x^{jn})|} \le \psi(x,x)/\tau .$$

This proves that the series (5.4.6) are uniformly and absolutely convergent. Hence Theorem 5.4.2 implies the statement of the theorem in the case that $x = y$. It then follows from the Cauchy-Schwarz inequality that the series (5.6.4) is absolutely convergent for all $x,y \in H$. Hence the right-hand side of equation (5.6.4) defines a sesquilinear form on H which agrees with φ_0 because the associated quadratic forms coincide.□

Using (5.3.1), we can rewrite the expansion formula (5.6.4) in the form

$$\varphi_0(x,y) = \sum_{j\in J} \frac{1}{\mu\lambda^j} \frac{\psi(x,x^j)\,\psi(x^j,y)}{\psi(x^j,x^j)} \quad , \quad x,y \in H . \tag{5.6.8}$$

It should be noticed that $\mu\lambda^j$ is nonzero for every $j \in J$. The above equation simplifies if we normalize the eigensystem x^j such that $\psi(x^j,x^j) = 1$, $j \in J$.

5.7 The Γ - operator

In this section we assume again that the eigenvalue problem (5.1.1) is left definite with respect to $\mu_1,\ldots,\mu_k$. We wish to show that formula (5.6.8) is closely related to the spectral representation of a compact Hermitian operator on a Hilbert space. We first need two lemmas.

LEMMA 5.7.1. *For every positive* τ *, the set of all* $j \in J$ *satisfying* $|\mu\lambda^j| \le \tau$ *is finite, where* $\mu = (0,\mu_1,\ldots,\mu_k)$.

The proof follows from Lemma 2.2.1 and Theorem 5.6.3.

LEMMA 5.7.2. *The sesquilinear form* φ_0 *is continuous with respect to the inner product* ψ *defined by (5.6.1).*

Proof. By (5.6.8), we have

$$|\varphi_0(x,x)| \le \sum_{j\in J} \frac{1}{|\mu\lambda^j|}\,\frac{|\psi(x,x^j)|^2}{\psi(x^j,x^j)} \quad \text{for all } x \in H , \tag{5.7.1}$$

where x^j , $j \in J$, is an eigensystem of (5.1.1). By Lemma 5.7.1, there is a positive constant c such that

$$1/|\mu\lambda^j| \le c \quad \text{for all } j \in J .$$

Hence (5.7.1) and Bessel's inequality yield

$$|\varphi_0(x,x)| \le c\,\psi(x,x) \quad \text{for all } x \in H .$$

It is well known [Kato(1966), Chapter I, (6.3.3)] that this inequality implies continuity of φ_0 with respect to ψ .□

By the above lemma, the sesquilinear form φ_0 can be continuously extended on the completion $(\bar{H},\bar{\psi})$ of the inner product space (H,ψ) . Therefore there is a uniquely determined bounded linear operator Γ on the Hilbert space $(\bar{H},\bar{\psi})$ satisfying

$$\bar{\psi}(\Gamma x,y) = \varphi_0(x,y) \quad \text{for all } x,y \in H . \tag{5.7.2}$$

Now let x^j , $j \in J$, be an eigensystem of (5.1.1) such that $\psi(x^j,x^j) = 1$ for every $j \in J$. Then this is an orthonormal system in $(\bar{H}, \bar{\psi})$. By (5.7.2), (5.6.8), we have

$$\bar{\psi}(\Gamma x,y) = \sum_{j \in J} (1/\mu\lambda^j)\, \bar{\psi}(x,x^j)\, \bar{\psi}(x^j,y) \quad \text{for all } x,y \in H . \tag{5.7.3}$$

Both sides of this equation represent continuous forms on $(\bar{H}, \bar{\psi})$. The continuity of the right hand side follows from the Cauchy-Schwarz inequality and the fact that the numbers $1/\mu\lambda^j$ converge to 0 as $|j| \to \infty$ according to Lemma 5.7.1. Hence (5.7.3) holds for all $x,y \in \bar{H}$ which proves that

$$\Gamma x = \sum_{j \in J} (1/\mu\lambda^j)\, \bar{\psi}(x,x^j)\, x^j \quad \text{for all } x \in \bar{H} . \tag{5.7.4}$$

It follows that Γ is a compact Hermitian operator on $(\bar{H}, \bar{\psi})$ with the spectral representation (5.7.4) i. e. the nonzero eigenvalues of Γ are given by $1/\mu\lambda^j$, $j \in J$, and the eigenspace of Γ associated with the nonzero eigenvalue $1/\tau$ is spanned by the vectors x^j for which j satisfies $\mu\lambda^j = \tau$. Every vector which lies in the orthocomplement of the kernel of Γ with respect to $(\bar{H}, \bar{\psi})$ can be expanded into a Fourier series of the eigensystem x^j , $j \in J$, Let us restate our results in the following

THEOREM 5.7.3. *Let the eigenvalue problem (5.1.1) be left definite. Then the operator* Γ *defined by (5.7.2) is compact and Hermitian on the Hilbert space* $(\bar{H}, \bar{\psi})$. *The nonzero eigenvalues of* Γ *are given by* $1/\mu\lambda$ *where* λ *runs through the set of eigenvalues of problem (5.1.1). The eigenspace of* Γ *associated with an eigenvalue* $1/\tau$ *is the direct sum of the eigenspaces of (5.1.1) associated with the finite number of eigenvalues* λ *satisfying* $\mu\lambda = \tau$.

5.8 The expansion theorem for right definite problems

In this section we assume that the eigenvalue problem (5.1.1) is right definite:

$$\varphi_0(x,x) > 0 \quad \text{for all nonzero decomposable tensors } x . \tag{5.8.1}$$

It follows from Theorem 5.2.1 that φ_0 is an inner product in H . By Theorem 2.7.1, the set J consists of all signed indices $(i,1)$ where i satisfies (2.2.3). Let x^j , $j \in J$, be an eigensystem of (5.1.1) such that

$$\varphi_0(x^j,x^j) = 1 \quad , \quad j \in J \quad . \tag{5.8.2}$$

Then x^j , $j \in J$, is an orthonormal system in (H , φ_0) .

As a corollary of Theorem 5.6.4 we have

THEOREM 5.8.1. *If the right definite problem (5.1.1) satisfies condition (2.7.2) for suitable* $\mu_1,\ldots,\mu_k$ *then* x^j , $j \in J$, *is an orthonormal basis of the inner product space* (H , φ_0) .

Proof. By Lemma 2.7.3, we can assume, without loss of generality, that (5.1.1) is right and left definite, simultaneously. Hence, by Theorem 5.6.4 and (5.8.2), there holds Parseval's identity

$$\varphi_0(x,x) = \sum_{j\in J} |\varphi_0(x,x^j)|^2 \quad \text{for all} \quad x \in H \quad .$$

This proves that x^j , $j \in J$, is an orthonormal basis of (H , φ_0) .□

There is an interesting special case of the above theorem. Assume that the eigenvalue problem (5.1.1) has the property that all k by k matrices

$$\langle A_{rs} u_r , u_r \rangle_r \; , \quad r,s = 1,\ldots,k \quad , \; u_r \in U_r \; , \tag{5.8.3}$$

are nonsingular M-matrices i. e. the matrices (5.8.3) have nonpositive offdiagonal elements and satisfy the equivalent conditions (i), (ii), (iii) of Lemma 5.5.1. Then, by property (ii), these matrices have positve determinants. Hence the eigenvalue problem is right definite. The determinants of the matrices (5.8.3) with ℓ^{th} row replaced by $(\mu_1,\ldots,\mu_k) = (1,\ldots,1)$ are positive, too. This follows from the properties (ii), (iii) if we expand the determinant according to the ℓth row. Hence the assumptions of Theorem 5.8.1 are satisfied, and therefore x^j , $j \in J$, is an orthonormal basis of (H , φ_0) .

If the eigenvalue problem (5.1.1) is right definite but not left definite then we can prove the following weaker result.

THEOREM 5.8.2. *Let the eigenvalue problem (5.1.1) be right definite, and let* $x \in H$ *be such that the linear functionals* $\varphi_s(.,x) = \overline{\varphi_s(x,.)}$ *are continuous on* (H , φ_0) *for every* $s = 1,\ldots,k$. *Then there holds the Fourier expansion*

$$x = \sum_{j \in J} \varphi_0(x, x^j)\, x^j \quad .$$

Proof. We choose an approximation (5.4.1) of the eigenvalue problem (5.1.1) which satisfies the conditions 1), 2), 3) of Section 5.4 such that $x \in H^1$. Further, for each n , let x^{jn} , $j \in J^n$, be an eigensystem of (5.4.1) which is related to the given eigensystem x^j , $j \in J$, according to (5.4.4).

The Riesz representation theorem for continuous linear functionals on a Hilbert space shows that there are vectors y_s , $s = 1,\dots,k$, lying in the completion $(\overline{H}, \overline{\varphi}_0)$ of (H, φ_0) such that

$$\varphi_s(x,y) = \overline{\varphi}_0(y_s, y) \quad , s = 1,\dots,k \quad , y \in H \quad . \tag{5.8.4}$$

Now Bessel's inequality gives

$$\sum_{j \in J^n} \frac{|\overline{\varphi}_0(y_s, x^{jn})|^2}{\varphi_0(x^{jn}, x^{jn})} \leq \overline{\varphi}_0(y_s, y_s) \quad .$$

Hence, by (5.8.4) and (5.3.1),

$$\sum_{j \in J^n} |\lambda_s^{jn}|^2 \frac{|\varphi_0(x, x^{jn})|^2}{\varphi_0(x^{jn}, x^{jn})} \leq \overline{\varphi}_0(y_s, y_s) \quad \text{for } s = 1,\dots,k \quad . \tag{5.8.5}$$

Lemma 5.4.1 shows that, for every positive τ , there is a positive integer m such that

$$\sum_{s=1}^{k} |\lambda_s^{jn}|^2 \geq \tau \quad \text{whenever } j \in J^n \ , \ |j| \geq m \quad .$$

Hence (5.8.5) shows that

$$\sum_{\substack{j \in J^n \\ |j| \geq m}} \frac{|\varphi_0(x, x^{jn})|^2}{\varphi_0(x^{jn}, x^{jn})} \leq (1/\tau) \sum_{s=1}^{k} \overline{\varphi}_0(y_s, y_s) \quad .$$

This proves that the series (5.4.6) are uniformly and absolutely convergent. Hence, by Theorem 5.4.2, the equation (5.4.8) holds which completes the proof of the theorem. □

To make clear the relation of this proof to that of Theorem 5.6.4, let us assume for the moment that the eigenvalue problem (5.1.1) is definite with respect to μ . Then both proofs depend on Bessel's inequality

$$\sum_{j\in J^n} \frac{|\psi(x,x^{jn})|^2}{\psi(x^{jn},x^{jn})} \leq \psi(x,x) \quad , \quad x \in H \quad , \tag{5.8.6}$$

where ψ is the inner product defined by

$$\psi(x,y) := \mu\varphi(x,y) \quad , \quad x,y \in H \quad .$$

Since $\psi(y,x^{jn}) = (\mu\lambda^{jn})\,\varphi_0(y,x^{jn})$, (5.8.6) yields

$$\sum_{j\in J^n} |\mu\lambda^{jn}| \frac{|\varphi_0(x,x^{jn})|^2}{|\varphi_0(x^{jn},x^{jn})|} \leq \psi(x,x) \quad , \quad x \in H \quad . \tag{5.8.7}$$

In the left definite case (5.8.7) implies the uniform and absolute convergence of the series (5.4.6) because $|\mu\lambda^{jn}|$ converges to infinity as $n \to \infty$. In the right definite case $\mu = (1,0,\ldots,0)$ this argument does not work because $\mu\lambda^{jn}$ is equal to 1 for all $j \in J^n$. Therefore Bessel's inequality (5.8.6) is not applied to the given vector x to be expanded but to the auxiliary vectors y_s associated with x . Demanding the existence of these auxiliary vectors, it is possible to generalize Theorem 5.8.2 to definite problems. However, this expansion theorem for definite problems is not very impressive because it does not include Theorem 5.6.4 as a special case.

The next theorem gives a sufficient condition for the eigensystem x^j , $j \in J$, to be an orthonormal basis of (H,φ_0) .

THEOREM 5.8.3. *For every* $r = 1,\ldots,k$, *let* B_r *be a positive definite bounded linear operator on* H_r *such that every of the following* $2k^2$ *eigenvalue problems*

$$\sum_{s=0}^{k} \lambda_s A'_{rs} x_r = 0, \quad \lambda_0 = 1, \ 0 \neq x_r \in H_r \ , \ r = 1,\ldots,k \quad ,$$

where

$$A'_{rs} = \begin{cases} A_{rs} & \text{if } (r,s) \neq (\ell,t) \ , \\ A_{rs} \pm B_r & \text{if } (r,s) = (\ell,t) \ , \end{cases}$$

and $\ell,t = 1,\ldots,k$, *is right definite. Then* x^j , $j \in J$, *is an orthonormal basis of the inner product space* (H,φ_0) .

Proof. Let $x_\ell = \langle B_\ell \cdot,\cdot\rangle_\ell$, and let $(-1)^{\ell+t}\varphi_{\ell t}$ denote the form determinant of the $k-1$ by $k-1$ array

$$\langle A_{rs}\cdot,\cdot\rangle_r \ , \quad r,s = 1,\dots,k \ , \ r \neq \ell \ , \ s \neq t \ .$$

Then the assumption of the theorem means that $(\varphi_0 \pm x_\ell \otimes \varphi_{\ell t})(x,x)$ is positive for all nonzero decomposable tensors $x \in H$ and all $\ell,t = 1,\dots,k$. By Theorem 5.2.1, this also holds for all nonzero tensors $x \in H$. It follows that

$$|(x_\ell \otimes \varphi_{\ell t})(x,x)| < \varphi_0(x,x) \ , \quad 0 \neq x \in H \ .$$

In particular, the eigenvalue problem (5.1.1) itself is right definite, and the sesquilinear forms $x_\ell \otimes \varphi_{\ell t}$, $\ell,t = 1,\dots,k$, are continuous on (H,φ_0) .

For decomposable tensors $x = x_1 \otimes \dots \otimes x_k$, $y = y_1 \otimes \dots \otimes y_k$ and $t = 1,\dots,k$, we have

$$-\varphi_t(x,y) = \sum_{\ell=1}^{k} \langle A_{\ell o}x_\ell, y_\ell\rangle_\ell \, \varphi_{\ell t}(\hat{x}_\ell,\hat{y}_\ell) \ ,$$

where $\hat{x}_\ell = x_1 \otimes \dots \otimes x_{\ell-1} \otimes x_{\ell+1} \otimes \dots \otimes x_k$. Now let $A_{\ell o}x_\ell$, $\ell = 1,\dots,k$, be of the form

$$A_{\ell o}x_\ell = B_\ell \tilde{x}_\ell \quad \text{for suitable } \tilde{x}_\ell \in H_\ell \ . \tag{5.8.8}$$

Then we obtain

$$-\varphi_t(x,y) = \sum_{\ell=1}^{k} x_\ell(\tilde{x}_\ell,y_\ell)\, \varphi_{\ell t}(\hat{x}_\ell,\hat{y}_\ell)$$

which can be written as

$$-\varphi_t(x,y) = \sum_{\ell=1}^{k} (x_\ell \otimes \varphi_{\ell t})\,(z_\ell,y) \ , \tag{5.8.9}$$

where $z_\ell = x_1 \otimes \dots \otimes x_{\ell-1} \otimes \tilde{x}_\ell \otimes x_{\ell+1} \otimes \dots \otimes x_k$. Since (5.8.9) holds for all decomposable y , this identity remains valid for all tensors y . We have already remarked that $x_\ell \otimes \varphi_{\ell t}$ is continuous on (H,φ_0) . Hence, by (5.8.9), $\varphi_t(x,.)$ is continuous on (H,φ_0) for every $t = 1,\dots,k$. By Theorem 5.8.2, x lies in the φ_0-closure of the linear hull of the system x^j , $j \in J$. This holds for all tensors $x \in F_1 \otimes \dots \otimes F_k$, where F_ℓ denotes the space of those

x_ℓ satisfying (5.8.8). Hence it remains to show that $F_1 \otimes \ldots \otimes F_k$ is dense in (H, φ_0).

To simplify the final argument, we assume, without loss of generality, that the operators $A_{\ell 0}$ are boundedly invertible for each ℓ. This is possible by virtue of Lemma 2.7.3. Then F_ℓ is dense in H_ℓ because the range of B_ℓ is dense. It follows that $F_1 \otimes \ldots \otimes F_k$ is dense in (H, φ_0) because the canonical map (5.2.1) is continuous as a map into (H, φ_0). □

5.9 Multiparameter eigenvalue problems for integral operators

As a special case of (5.1.1), we consider the eigenvalue problem for integral operators

$$-x_r(\xi_r) + \sum_{s=1}^{k} \lambda_s \int_{a_r}^{b_r} a_{rs}(\xi_r, \eta_r)\, x_r(\eta_r)\, d\eta_r = 0,\ 0 \neq x_r \in H_r,\ r = 1,\ldots,k. \quad (5.9.1)$$

The space $H_r = L^2(a_r, b_r)$, $-\infty < a_r < b_r < \infty$, is endowed with the usual inner product $<\ ,\ >_r$. The kernels a_{rs} of the Fredholm integral operators

$$(A_{rs}\, x_r)(\xi_r) = \int_{a_r}^{b_r} a_{rs}(\xi_r, \eta_r)\, x_r(\eta_r)\, d\eta_r\ ,\quad r,s = 1,\ldots,k\ , \quad (5.9.2)$$

are continuous and complex-valued on $[a_r, b_r] \times [a_r, b_r]$ and satisfy

$$a_{rs}(\xi_r, \eta_r) = \overline{a_{rs}(\eta_r, \xi_r)} \quad \text{for } \xi_r, \eta_r \in [a_r, b_r]\ .$$

It is well known that these integral operators are compact and Hermitian on H_r. Hence the results of Chapters 2 and 5 can be applied to the eigenvalue problem (5.9.1). To simplify the notation, we have set $A_r = 0$ in (5.9.1) but the following results remain true apart from some obvious changes if we include Fredholm integral operators A_r in the formulation of the eigenvalue problem (5.9.1). We remark that the solutions x_r of (5.9.1) are continuous functions on $[a_r, b_r]$ because the ranges of the integral operators (5.9.2) consist of continuous functions only.

For $x_r \in H_r$, $r = 1,\ldots,k$, we identify the decomposable tensor $x = x_1 \otimes \ldots \otimes x_k$ with the function

$$x(\xi) = \prod_{r=1}^{k} x_r(\xi_r)\ ,\quad \xi = (\xi_1,\ldots,\xi_k) \in \Pi := \prod_{r=1}^{k} [a_r, b_r]$$

lying in $L^2(\Pi)$. Then the tensor product $H = H_1 \otimes \ldots \otimes H_k$ is the linear hull of these functions. The tensorial product of the inner products $<\ ,\ >_r$ in H_r is equal to the usual inner product $<\ ,\ >$ in $L^2(\Pi)$ restricted to H . Since H is dense in $(L^2(\Pi),<\ ,\ >)$, we obtain that $(L^2(\Pi),<\ ,\ >)$ is the completion of H with respect to $<\ ,\ >$.

Let us now define a continuous function d_o on $\Pi \times \Pi$ by

$$d_o(\xi,\eta) := \det_{1 \le r,s \le k} a_{rs}(\xi_r,\eta_r) \quad , \tag{5.9.3}$$

and an integral operator Δ_o on $L^2(\Pi)$ by

$$(\Delta_o x)(\xi) := \int_\Pi d_o(\xi,\eta)\, x(\eta)\, d\eta \quad .$$

We note that

$$\varphi_o(x,y) = <\Delta_o x,y> \quad \text{for} \quad x,y \in H \quad . \tag{5.9.4}$$

If the eigenvalue problem (5.9.1) is right definite then $\varphi_o(x,x)$ is positive for all nonzero $x \in H$. Hence, by (5.9.4), Δ_o is a positive semidefinite integral operator. Positive semidefinite integral operators satisfy the following useful inequality.

LEMMA 5.9.1. *Let* g *be a continuous function on* $\Pi \times \Pi$ *which satisfies* $g(\xi,\eta) = \overline{g(\eta,\xi)}$ *for* $\xi,\eta \in \Pi$ *. Assume that the integral operator with kernel* g ,

$$(\Phi x)(\xi) = \int_\Pi g(\xi,\eta)\, x(\eta)\, d\eta \quad ,$$

is positive semidefinite i. e.

$$<\Phi x,x> \ \ge 0 \quad \textit{for all} \quad x \in L^2(\Pi) \quad .$$

Then

$$|(\Phi x)(\xi)|^2 \le g(\xi,\xi) <\Phi x,x> \quad \textit{for} \quad x \in L^2(\Pi)\ ,\ \xi \in \Pi \quad . \tag{5.9.5}$$

Proof. The Cauchy-Schwarz inequality applied to the positive semidefinite form $<\Phi\ .\ ,\ .\ >$ yields

$$|<\Phi x,y>|^2 \le <\Phi x,x> <\Phi y,y> \quad \text{for all} \quad x,y \in L^2(\Pi) \quad . \tag{5.9.6}$$

For fixed x and $\tilde{\xi} \in \Pi$, let y run through a sequence y^n of continuous nonnegative functions satisfying

$$\int_\Pi y^n(\eta)\, d\eta = 1 \quad \text{and} \quad y^n(\xi) = 0 \ \text{ if } \ \|\xi - \tilde{\xi}\| \geq 1/n \ .$$

Some elementary analysis shows that

$$\langle \Phi x, y^n \rangle \to (\Phi x)(\tilde{\xi}) \ , \quad \langle \Phi y^n, y^n \rangle \to g(\tilde{\xi},\tilde{\xi}) \quad \text{as} \quad n \to \infty \ .$$

Hence (5.9.6) gives (5.9.5) with $\tilde{\xi}$ in place of ξ .□

The next lemma follows from the above lemma applied to $g = d_0$ and the fact that the continuous function d_0 is bounded on Π .

LEMMA 5.9.2. *If (5.9.1) is right definite then* Δ_0 *is a continuous linear operator from* (H, φ_0) *into* $(C(\Pi), \|\ \|_\infty)$ *, where* $C(\Pi)$ *is the linear space of continuous complex-valued functions on* Π *and* $\|\ \|_\infty$ *denotes the maximum norm* .

Using Lemma 5.9.2 we can prove

THEOREM 5.9.3. *Let* x^j, $j \in J$ *, be an eigensystem of the right definite eigenvalue problem (5.9.1) such that* $\varphi_0(x^j,x^j) = 1$, $j \in J$ *. Then, if* $x \in H$ *satisfies*

$$x = \sum_{j \in J} \varphi_0(x,x^j)\, x^j \ , \tag{5.9.7}$$

the function $\Delta_0 x$ *can be expanded into the series*

$$\Delta_0 x = \sum_{j \in J} \varphi_0(x,x^j)\, \Delta_0 x^j \ , \tag{5.9.8}$$

which is uniformly and absolutely convergent i. e. for every positive ε *there is a positive integer* m *such that*

$$\sum_{\substack{j \in J \\ |j| \geq m}} |\varphi_0(x,x^j)(\Delta_0 x^j)(\xi)| < \varepsilon \quad \textit{for all} \quad \xi \in \Pi \ .$$

Proof. The series (5.9.7) converges unconditionally to x with respect to φ_0 . Hence Lemma 5.9.2 shows that the series (5.9.8) converges uniformly to $\Delta_0 x$ independently of the order of the summation.

It remains to show that the series (5.9.8) is uniformly and absolutely convergent. By Lemma 5.9.2, there is a constant c such that

$$\|\Delta_0 y\|_\infty^2 \le c\,\varphi_0(y,y) \quad \text{for all} \quad y \in H \ . \tag{5.9.9}$$

For a finite subset K of J and $\xi \in \Pi$, we set

$$z := \sum_{j\in K} \alpha_j x^j \ , \text{ where } \bar{\alpha}_j := (\Delta_0 x^j)(\xi) \ .$$

Since x^j, $j \in J$, is orthonormal with respect to φ_0, we then have

$$(\Delta_0 z)(\xi) = \sum_{j\in K} |\alpha_j|^2 = \varphi_0(z,z) \ .$$

Hence, by (5.9.9),

$$\varphi_0(z,z)^2 = (\Delta_0 z)(\xi)^2 \le c\,\varphi_0(z,z) \ ,$$

consequently

$$\sum_{j\in K} |(\Delta_0 x^j)(\xi)|^2 \le c \ .$$

It follows from the Cauchy-Schwarz inequality that

$$\Big(\sum_{j\in K} |\varphi_0(x,x^j)\,(\Delta_0 x^j)(\xi)|\Big)^2 \le c \sum_{j\in K} |\varphi_0(x,x^j)|^2 \ . \tag{5.9.10}$$

This proves the uniform and absolute convergence of the series (5.9.8) because

$$\sum_{j\in J} |\varphi_0(x,x^j)|^2 \le \varphi_0(x,x)^2 < \infty \ . \square$$

The validity of (5.9.7) has to be shown by use of the results of Section 5.8. The above theorem only states that if x can be expanded into the Fourier series (5.9.7) then we can improve the convergence of this series by applying the operator Δ_0 term by term.

We now turn to the left definite case. Let us assume that the eigenvalue problem (5.9.1) is left definite with repect to $\mu_1,\dots,\mu_k$. For $\xi = (\xi_1,\dots,\xi_k)$, $\eta = (\eta_1,\dots,\eta_k)$ and $\ell = 1,\dots,k$, let $g_\ell(\hat{\xi}_\ell,\hat{\eta}_\ell)$ denote the determinant of the k by k matrix $a_{rs}(\xi_r,\eta_r)$, $r,s = 1,\dots,k$, with ℓ^{th} row replaced by $(\mu_1,\dots,\mu_k)$. We have used the abbreviation

$$\hat{\xi}_\ell = (\xi_1,\ldots,\xi_{\ell-1},\xi_{\ell+1},\ldots,\xi_k) \in \hat{\Pi}_\ell := \prod_{\substack{r=1\\ r\neq\ell}}^{k} [a_r,b_r] \quad .$$

The integral operator with the kernel function g_ℓ is denoted by Φ_ℓ. It follows from the the assumed left definiteness and Theorem 5.2.1 that Φ_ℓ is a positive semidefinite integral operator on $L^2(\hat{\Pi}_\ell)$. Hence we can apply Lemma 5.9.1 to this operator to obtain

$$|(\Phi_\ell x)(\hat{\xi}_\ell)|^2 \leq g_\ell(\hat{\xi}_\ell,\hat{\xi}_\ell) \langle\Phi_\ell x,x\rangle \ , \ x \in L^2(\hat{\Pi}_\ell) \ , \ \hat{\xi}_\ell \in \hat{\Pi}_\ell \quad . \tag{5.9.11}$$

The inner product ψ used in Section 5.6 can now be written as

$$\psi(x,y) = \sum_{\ell=1}^{k} \int_{a_\ell}^{b_\ell} \int_{\hat{\Pi}_\ell} \int_{\hat{\Pi}_\ell} g_\ell(\hat{\alpha}_\ell,\hat{\beta}_\ell)\ x(\hat{\beta}_\ell;\eta_\ell)\ \overline{y(\hat{\alpha}_\ell;\eta_\ell)}\ d\hat{\alpha}_\ell\ d\hat{\beta}_\ell\ d\eta_\ell \ , \tag{5.9.12}$$

where, for abbreviation,

$$y(\hat{\alpha}_\ell;\eta_\ell) = y(\alpha_1,\ldots,\alpha_{\ell-1},\eta_\ell,\alpha_{\ell+1},\ldots,\alpha_k) \quad .$$

To prove (5.9.12), we first note that it suffices to assume that $x = x_1 \otimes \ldots \otimes x_k$ and $y = y_1 \otimes \ldots \otimes y_k$ are decomposable. In this case we obtain (5.9.12) by expanding the determinant

$$\psi(x,y) = \det \begin{pmatrix} 0 & \mu_1 & \cdots & \mu_k \\ -\langle x_1,y_1\rangle_1 & \langle A_{11}x_1,y_1\rangle_1 & \cdots & \langle A_{1k}x_1,y_1\rangle_1 \\ \vdots & \vdots & & \vdots \\ -\langle x_k,y_k\rangle_k & \langle A_{k1}x_k,y_k\rangle_k & \cdots & \langle A_{kk}x_k,y_k\rangle_k \end{pmatrix}$$

according to the zeroth column.

Analogously to Lemma 5.9.2, we have

LEMMA 5.9.4. *If (5.9.1) is left definite then* Δ_0 *is a continuous linear operator from* (H,ψ) *into* $(C(\Pi),\|\ \|_\infty)$.

Proof. To simplify the notation, we assume that $(\mu_1,\ldots,\mu_k) = (0,\ldots,0,1)$. The proof is analogous in the general case. Expanding the determinant (5.9.3) according to the last column, we obtain

$$d_0(\xi,\eta) = \sum_{\ell=1}^{k} a_{\ell k}(\xi_\ell,\eta_\ell)\ g_\ell(\hat{\xi}_\ell,\hat{\eta}_\ell) \quad .$$

Hence, for all $x \in L^2(\Pi)$,

$$(\Delta_0 x)(\xi) = \sum_{\ell=1}^{k} \int_{a_\ell}^{b_\ell} a_{\ell k}(\xi_\ell,\eta_\ell) \int_{\hat{\Pi}_\ell} g_\ell(\hat{\xi}_\ell,\hat{\eta}_\ell)\, x(\eta)\, d\hat{\eta}_\ell\, d\eta_\ell \ .$$

The estimate (5.9.11) shows that

$$|(\Delta_0 x)(\xi)| \le \sum_{\ell=1}^{k} \int_{a_\ell}^{b_\ell} |a_{\ell k}(\xi_\ell,\eta_\ell)|\, g_\ell(\hat{\xi}_\ell,\hat{\xi}_\ell)^{1/2} \cdot$$

$$\cdot \left(\int_{\hat{\Pi}_\ell} \int_{\hat{\Pi}_\ell} g_\ell(\hat{\alpha}_\ell,\hat{\beta}_\ell)\, x(\hat{\beta}_\ell;\eta_\ell)\, \overline{x(\hat{\alpha}_\ell;\eta_\ell)}\, d\hat{\alpha}_\ell\, d\hat{\beta}_\ell \right)^{1/2} d\eta_\ell \ .$$

The Cauchy-Schwarz inequality applied to the integrals with limits a_ℓ and b_ℓ and then to the sum gives

$$|(\Delta_0 x)(\xi)|^2 \le c(\xi)\, \psi(x,x) \quad , \quad x \in H \ ,$$

where

$$c(\xi) = \sum_{\ell=1}^{k} g_\ell(\hat{\xi}_\ell,\hat{\xi}_\ell) \int_{a_\ell}^{b_\ell} |a_{\ell k}(\xi_\ell,\eta_\ell)|^2\, d\eta_\ell \ .$$

This shows that

$$\| \Delta_0 x \|_\infty^2 \le \| c \|_\infty\, \psi(x,x) \quad \text{for all} \quad x \in H$$

which proves the lemma.□

This lemma together with Theorem 5.6.4 yields the following expansion theorem.

THEOREM 5.9.5. *Let* x^j, $j \in J$, *be an eigensystem of the left definite eigenvalue problem (5.9.1) such that* $\psi(x^j,x^j) = 1$, $j \in J$. *Then every function* $\Delta_0 x$, $x \in H$, *can be expanded into the uniformly and absolutely convergent series*

$$\Delta_0 x = \sum_{j \in J} \psi(x,x^j)\, \Delta_0 x^j \ . \tag{5.9.13}$$

Proof. The uniform and absolute convergence of the series (5.9.13) can be shown as in the proof of Theorem 5.9.3. We just have to replace φ_0 by ψ and Lemma 5.9.2 by Lemma 5.9.4. It remains to prove that equation (5.9.13) holds for all $x \in H$. We know from Theorem 5.6.4 that

$$\varphi_0(x,y) = \sum_{j\in J} \frac{\varphi_0(x,x^j)\ \varphi_0(x^j,y)}{\varphi_0(x^j,x^j)} \quad \text{for all} \quad x,y \in H \ . \tag{5.9.14}$$

It follows from (5.3.1) and $\psi(x^j,x^j) = 1$ that

$$\psi(x,x^j) = \varphi_0(x,x^j) \ / \ \varphi_0(x^j,x^j) \ .$$

Hence, by (5.9.14) and (5.9.4),

$$<\Delta_0 x - \sum_{j\in J} \psi(x,x^j)\ \Delta_0 x^j \ , y> \ = \ 0 \ .$$

This proves (5.9.13) because y is arbitrary.□

We remark that the statement of the above theorem also holds for functions $\Delta_0 x$ with $x \in L^2(\Pi)$ if we extend the sesquilinear form ψ continuously on $L^2(\Pi)$ using (5.9.12). Then ψ is positive semidefinite on $L^2(\Pi)$. The proof of the uniform and absolute convergence of the series (5.9.13) for $x \in L^2(\Pi)$ is the same as for $x \in H$. Using the inequality (5.9.10) with ψ in place of φ_0 , we see that the right-hand side of equation (5.9.13) defines a continuous linear operator from $(L^2(\Pi), \psi)$ into $(C(\Pi), \|\ \|_\infty)$. Since the same is true for the left-hand side, equation (5.9.13) extends to all $x \in L^2(\Pi)$.

5.10 Notes for Chapter 5

It is usual in multiparameter spectral theory to complete the tensor product H with respect to the tensorial product $<\ ,\ >$ of the inner products $<\ ,\ >_r$ in order to work within a Hilbert space. It should be noticed that we did not use this completion apart from Section 5.9. In fact, the analysis of this chapter shows that the sesquilinear forms φ_s but not $<\ ,\ >$ play the crucial role in the study of the eigenvalue problem (5.1.1).

The method of approximation of multiparameter eigenvalue problems by finite dimensional problems was applied by Carmichael (1921a), (1921b), (1922) to eigenvalue problems of different types. Atkinson (1972) used this method in Chapter 11 of his book in the case of right definite problems. The approximation method can also be found in papers of Binding, Källström and Sleeman (1982) and Binding (1984c). However, the particular choice of the approximation scheme explained in Section 5.4 is new.

The theory of left definite problems presented in Sections 5.5, 5.6, 5.7 is given here for the first time. Theorem 5.7.3 shows that, in the left definite case, every vector in $(\overline{H}, \overline{\psi})$ which is orthogonal to the kernel of Γ can be expanded into a series of eigenvectors. Binding, Källström and Sleeman (1982) proved a similar theorem but there are additional conditions for a vector to be expandable; see Theorem 4.2 of that paper. Our results show that these additional conditions are superfluous in the left definite case.

Pell (1922) investigated the two-parameter eigenvalue problem

$$-x_1 + \lambda_1 A_{11}x_1 + \lambda_2 A_{12}x_1 = 0 , \; 0 \neq x_1 \in H_1 ,$$

$$-x_2 + \lambda_1 A_{21}x_2 + \lambda_2 A_{22}x_2 = 0 , \; 0 \neq x_2 \in H_2 .$$

The spaces H_1 and H_2 are sequence spaces ℓ^2 . The operators A_{rs} are given by infinite real symmetric Hilbert-Schmidt matrices. The operators A_{12} and $-A_{22}$ are assumed to be positive definite. Hence the eigenvalue problem is left definite with respect to $\mu_1 = -1$, $\mu_2 = 0$. Theorem 1 of Pell's paper states that the eigenvalue problem under consideration has at least one eigenvalue if φ_0 is not identically zero. We note that this statement is a special case of Theorem 2.5.3. Theorem 2 of Pell's paper establishes formula (5.6.8), where x and y are chosen as the canonical unit vectors of ℓ^2 . Finally, Theorem 4 of Pell's paper is a special case of Theorem 5.9.5. Hence the results of Pell are completely generalized to left definite problems involving more than two parameters. It should be mentioned that Pell's paper contains an error on page 200, line 10, which seems difficult to correct.

The limiting process in the right definite case was carried out by Atkinson (1972) and, in modified form, by Binding, Källström and Sleeman (1982) and Binding (1984c). Theorem 5.8.1 is new; the following special case was essentially shown by Volkmer (1987). The assumption of Theorem 5.8.2 is slightly weaker than that of the similar theorems of [Atkinson (1972), Theorem 1.10.1] and [Binding, Källström and Sleeman (1982), Theorem 5.1].

Surprisingly, it is an open question whether an eigensystem of a right definite eigenvalue problem is automatically complete in (H, φ_0) . In other words, it is not known whether the continuity assumption of Theorem 5.8.2 is really necessary. This assumption is clearly superfluous in the trivial case $k = 1$. Volkmer (1987) has shown the completeness of eigensystems for two-parameter right definite problems. The question is open for k greater than two. The sufficient condition for completeness of eigensystems given by Theorem 5.8.3 is taken from Volkmer (1984a), (1985).

The results of section 5.9 generalize those of Pell (1922).

CHAPTER 6

MULTIPARAMETER EXPANSION THEOREMS FOR UNBOUNDED OPERATORS

6.1 Introduction

We suppose given k nonzero separable Hilbert spaces $(H_r, < , >_r)$, $r = 1,\dots,k$. For each r, let A_{rs}, $s = 1,\dots,k$, be a set of k bounded Hermitian operators on H_r, and let $A_{ro} : H_r \supset D_r \to H_r$ be a selfadjoint operator, bounded above with compact resolvent.

We shall study multiparameter expansion theorems for the eigenvalue problem of Chapter 3

$$\sum_{s=0}^{k} \lambda_s A_{rs} x_r = 0 , \quad \lambda_0 = 1, 0 \neq x_r \in D_r, r = 1,\dots,k , \tag{6.1.1}$$

via the transformed problem of Section 3.3

$$\sum_{s=0}^{k} \lambda_s \tilde{A}_{rs} y_r = 0 , \quad \lambda_0 = 1, 0 \neq y_r \in H_r, r = 1,\dots,k , \tag{6.1.2}$$

where

$$A_{ro} = \gamma I_r - S_r^{-2} \quad \text{for } \gamma \text{ sufficiently large,}$$

$$\tilde{A}_{ro} := \gamma S_r^2 - I_r , \tilde{A}_{rs} := S_r A_{rs} S_r , \; r,s = 1,\dots,k .$$

After the preparatory Section 6.2, we relate the eigenspaces of (6.1.1) to those of (6.1.2). Then it will be easy to carry over the expansion theorems of Chapter 5 from problem (6.1.2) to the given problem (6.1.1); the left definite case and the right definite case are treated in Sections 6.4, 6.5, respectively. Special attention is paid to expansion theorems under strict definiteness conditions.

As an important special case, we then study expansion theorems for multiparameter boundary eigenvalue problems. In applications these problems arise from boundary eigenvalue problems for partial differential operators by the method of separation of variables; see Section 6.6. The left definite case and the right definte case are treated in Sections 6.7, 6.8, respectively. Under strict definiteness conditions, we show that the expansions converge in suitably chosen Sobolev spaces. As examples, we consider the expansion of functions in series of Lamé's wave functions and in series of Lamé products; see Sections 6.9 and 6.10, respectively. References to the literature are given in Section 6.11.

6.2 Forms and operators on tensor products

The following general results on tensor products will be useful, in particular, when we deal with strict definiteness conditions.

Let $(H_r, <\,,\,>_r)$, $r = 1,\dots,k$, be complex inner product spaces, not necessarily complete, and let

$$H_1 \times \dots \times H_k \ni (x_1,\dots,x_k) \mapsto x_1 \otimes \dots \otimes x_k \in H := H_1 \otimes \dots \otimes H_k \tag{6.2.1}$$

be the canonical map associated with their tensor product. We denote the tensorial product of the forms $<\,,\,>_r$, $r = 1,\dots,k$, by $<\,,\,>$. Hence

$$<x_1 \otimes \dots \otimes x_k , y_1 \otimes \dots \otimes y_k> = \prod_{r=1}^{k} <x_r,y_r>_r \quad \text{for} \quad x_r,y_r \in H_r \ . \tag{6.2.2}$$

We then have

LEMMA 6.2.1. *The form* $<\,,\,>$ *is an inner product on* H *making the map (6.2.1) continuous.*

Proof. For given $x \in H$, there are finite dimensional subspaces F_r of H_r such that $x \in F := F_1 \otimes \dots \otimes F_k$. For each r , we choose a basis $x_r^{m_r}$, $m_r = 1,\dots,\dim F_r$, of F_r which is orthonormal with respect to $<\,,\,>_r$. Then the decomposable tensors

$$x^m = x_1^{m_1} \otimes \dots \otimes x_k^{m_k} \ , \ m = (m_1,\dots,m_k) \ , \tag{6.2.3}$$

form a basis of F satisfying

$$<x^m,x^n> = \begin{cases} 1 & \text{if } m = n \ , \\ 0 & \text{if } m \neq n \ . \end{cases}$$

Hence if we expand x as

$$x = \sum_m \alpha_m x^m \tag{6.2.4}$$

then we obtain

$$<x,x> = \sum_m |\alpha_m|^2 .$$

This shows that $<x,x>$ is positive for all nonzero x. Continuity of the multilinear map (6.2.1) now follows from (6.2.2).□

We mention that positive definiteness of $< , >$ is also a consequence of Theorem 5.2.1 applied to diagonal arrays but the above proof is more direct.

The next lemma shows that the tensorial product of continuous Hermitian forms on $(H_r, < , >_r)$ is continuous on $(H, < , >)$.

LEMMA 6.2.2. *Let* ψ_r, $r = 1,\dots,k$, *be Hermitian sesquilinear forms on* H_r, *and let* $\psi = \psi_1 \otimes \dots \otimes \psi_k$ *be their tensorial product. Then*

$$|\psi_r(x_r,x_r)| \leq <x_r,x_r>_r , \quad x_r \in H_r, \ r = 1,\dots,k , \tag{6.2.5}$$

implies that

$$|\psi(x,x)| \leq <x,x> , \quad x \in H . \tag{6.2.6}$$

Proof. For a given $x \in H$, we choose spaces F_r as in the proof of the previous lemma. There are Hermitian operators B_r on F_r such that

$$\psi_r(x_r,y_r) = <B_r x_r,y_r>_r , \quad x_r,y_r \in F_r, \ r = 1,\dots,k .$$

Let $x_r^{m_r}$, $m_r = 1,\dots,\dim F_r$, be an orthonormal basis of $(F_r, < , >_r)$ consisting of eigenvectors of B_r. For the basis (6.2.3) of F we then have

$$\psi(x^m,x^n) = 0 \quad \text{if} \quad m \neq n ,$$

and, by (6.2.5),

$$|\psi(x^m,x^m)| \leq <x^m,x^m> = 1 \quad \text{for every} \quad m .$$

Hence if we expand x according to (6.2.4) we obtain

$$|\psi(x,x)| = |\sum_m |\alpha_m|^2 \psi(x^m,x^m)| \leq \sum_m |\alpha_m|^2 = <x,x> .$$

This proves (6.2.6).□

The following result is a supplement to Theorem 5.2.1.

THEOREM 6.2.3. *Let* ψ_{rs}, $r,s = 1,\dots,k$ *, be continuous Hermitian sesquilinear forms on* $(H_r, <\ ,\ >_r)$ *, and let* ψ *be their form determinant. Then the existence of a positive* ε *satisfying*

$$\psi(x,x) \geq \varepsilon <x,x> \quad \textit{for all decomposable} \quad x \tag{6.2.7}$$

implies the existence of a positive η *such that*

$$\psi(x,x) \geq \eta <x,x> \quad \textit{for all} \quad x \in H \ .$$

Proof. We proof the statement by induction on k. For $k = 1$, we can choose $\eta = \varepsilon$. Now assume that the statement is true for $k-1$ in place of k and let ψ_{rs}, $r,s = 1,\dots,k$, satisfy (6.2.7). We fix a nonzero $y_k \in H_k$. By performing elementary operations on the columns of the array ψ_{rs}, $r,s = 1,\dots,k$, we may assume, without loss of generality, that

$$\psi_{kk}(y_k,y_k) = 1 \ , \ \psi_{ks}(y_k,y_k) = 0 \quad \text{if} \quad s \neq k \ . \tag{6.2.8}$$

Now let ψ_0 be the form determinant of the array ψ_{rs}, $r,s = 1,\dots,k-1$. Then (6.2.8) and (6.2.7) show that, for every $y = y_1 \otimes \dots \otimes y_{k-1}$,

$$\psi_0(y,y) = \psi(y \otimes y_k \, , y \otimes y_k) \geq \varepsilon <y,y>_0 <y_k,y_k>_k \ ,$$

where $<\ ,\ >_0$ denotes the tensorial product of the forms $<\ ,\ >_r$, $r = 1,\dots,k-1$. It follows from the induction hypothesis that there is a positive η such that

$$\psi_0(y,y) \geq \eta <y,y>_0 \quad \text{for all} \quad y \in H_1 \otimes \dots \otimes H_{k-1} \ .$$

Hence, by Lemma 6.2.2 applied to the forms $\psi_1 = \eta <\ ,\ >_0$ on $(H_1 \otimes \dots \otimes H_{k-1}, \psi_0)$ and $\psi_2 = \chi := <\ ,\ >_k$ on $(H_k, <\ ,\ >_k)$, we obtain

$$(\psi_0 \otimes \chi)(x,x) \geq \eta <x,x> \quad \text{for all} \quad x \in H \ . \tag{6.2.9}$$

By continuity of the forms ψ_{rs}, there is a positive θ such that

$$\psi_0(y,y) \leq \theta <y,y>_0 \quad \text{for all decomposable} \quad y \ . \tag{6.2.10}$$

Now let $\tilde{\psi}$ be the form determinant of the array ψ_{rs}, $r,s = 1,\ldots,k$, with ψ_{kk} replaced by $\psi_{kk} - (\varepsilon/2\theta)\chi$. Then (6.2.7) and (6.2.10) show that, for all decomposable $x \in H$,

$$\tilde{\psi}(x,x) = \psi(x,x) - \frac{\varepsilon}{2\theta}(\psi_0 \otimes \chi)(x,x) \geq \frac{\varepsilon}{2}\langle x,x\rangle \quad .$$

Thus, by Theorem 5.2.1, the form $\tilde{\psi}$ is positive definite. Hence

$$\psi(x,x) \geq \frac{\varepsilon}{2\theta}(\psi_0 \otimes \chi)(x,x) \quad \text{for all} \quad x \in H \quad .$$

This together with (6.2.9) completes the proof.□

In the following sections we shall use the *tensorial product*

$$T = T_1 \otimes \ldots \otimes T_k : E_1 \otimes \ldots \otimes E_k \to F_1 \otimes \ldots \otimes F_k$$

of linear maps $T_r : E_r \to F_r$, $r = 1,\ldots,k$, where E_r and F_r are complex linear spaces. The linear map T is uniquely determined by

$$T(x_1 \otimes \ldots \otimes x_k) = T_1x_1 \otimes \ldots \otimes T_kx_k, \quad x_r \in E_r \quad .$$

We recall that the tensorial product of isomorphisms is again an isomorphism.

For an array of linear operators

$$T_{rs} : E_r \to F_r, \quad r,s = 1,\ldots,k \quad ,$$

we define its *operator determinant* T by

$$T = \sum_{\tau} \varepsilon_{\tau} T_{1\tau(1)} \otimes \ldots \otimes T_{k\tau(k)} \quad ,$$

where τ runs through permutations of $1,\ldots,k$ and ε_{τ} is 1 or -1 if τ is even or odd, respectively.

6.3 Transformation of eigenvectors

We consider the eigenvalue problem (6.1.1). If $\lambda = (1,\lambda_1,\ldots,\lambda_k)$ is an eigenvalue then we define the *eigenspace* $E(\lambda)$ to be the tensor product

$$E(\lambda) = E_1(\lambda) \otimes \ldots \otimes E_k(\lambda) \subset H = H_1 \otimes \ldots \otimes H_k \quad ,$$

where $E_r(\lambda)$ is the kernel of the operator

$$A_{ro} + \sum_{s=1}^{k} \lambda_s A_{rs} \quad .$$

The nonzero elements of $E(\lambda)$ are the *eigenvectors* belonging to the eigenvalue λ .

If we denote the eigenspaces of (6.1.2) by

$$\widetilde{E}(\lambda) = \widetilde{E}_1(\lambda) \otimes \ldots \otimes \widetilde{E}_k(\lambda)$$

then a simple calculation shows that

$$S_r(\widetilde{E}_r(\lambda)) = E_r(\lambda), \; r = 1,\ldots,k \quad . \tag{6.3.1}$$

Since S_r is one-to-one, $E_r(\lambda)$ and $\widetilde{E}_r(\lambda)$ have the same finite dimension. The relation (6.3.1) can be rewritten as

$$S(\widetilde{E}(\lambda)) = E(\lambda) \quad , \tag{6.3.2}$$

where $S := S_1 \otimes \ldots \otimes S_k$ is the tensorial product of the operators S_r . The map S_r is an isomorphism from H_r onto the range of S_r which was denoted by G_r . Hence S is an isomorphism from H onto $G := G_1 \otimes \ldots \otimes G_k$.

As in Section 4.3, we define the sesquilinear forms $(-1)^s \varphi_s$, $s = 0,\ldots,k$, as the form determinants of the k by k array (4.3.1). The form φ_0 is defined on H . The forms $\varphi_1,\ldots,\varphi_k$ are defined on G if we use the closures $\bar{\omega}_r$ of the forms $\omega_r(x_r,y_r) = \langle A_{ro}\, x_r,y_r\rangle_r$; see Section 3.2. The relation between

$$\varphi(x,y) = (\varphi_0(x,y),\ldots,\varphi_k(x,y))$$

and the corresponding vector $\widetilde{\varphi}(x,y)$ belonging to (6.1.2) is given by

$$\widetilde{\varphi}(x,y) = \varphi(S\,x\,,\,S\,y\,) \quad , \quad x,y \in H \quad . \tag{6.3.3}$$

It follows from (5.3.1), (6.3.2), (6.3.3) that

$$\varphi_s(x,y) = \lambda_s\, \varphi_0(x,y), \quad x \in E(\lambda), \quad y \in G, \quad s = 1,\ldots,k \; . \tag{6.3.4}$$

Of course, (6.3.4) can also be shown directly without using the results of Chapter 5. Similarly, we see that Lemma 5.3.1 remains true for the eigenvalue problem (6.1.1).

Analogously to the definition of the form determinants $(-1)^s \varphi_s$, let $(-1)^s \Delta_s$ denote the operator determinant of the array

$$A_{rt},\ r = 1,\dots,k,\ t = 0,\dots,k,\ t \neq s\ .$$

Then Δ_0 is a linear operator on H satisfying

$$\langle \Delta_0 x,y \rangle = \varphi_0(x,y) \quad \text{for} \quad x,y \in H\ , \tag{6.3.5}$$

where $\langle\ ,\ \rangle$ denotes the tensorial product of the forms $\langle\ ,\ \rangle_r$, $r = 1,\dots,k$. By Lemma 6.2.1, $\langle\ ,\ \rangle$ is an inner product in H . The definition (4.2.2) of form determinants and Lemma 6.2.2 show that φ_0 is a finite sum of sesquilinear forms which are continuous on $(H, \langle\ ,\ \rangle)$. Hence φ_0 is continuous, too. It follows from (6.3.5) that Δ_0 is continuous on $(H, \langle\ ,\ \rangle)$.

The operators $\Delta_1,\dots,\Delta_k$ are linear maps from $D := D_1 \otimes \dots \otimes D_k$ into H which satisfy

$$\langle \Delta_s x,y \rangle = \varphi_s(x,y)\ ,\quad x \in D,\ y \in G\ . \tag{6.3.6}$$

The forms $\varphi_1,\dots,\varphi_k$ are not continuous in general but (6.3.6) yields

LEMMA 6.3.1. *For every* $x \in D$ *and* $s = 1,\dots,k$, *the linear functional* $\varphi_s(\cdot\ ,x) = \overline{\varphi_s(x\ ,\cdot)}$ *is continuous on* $(G, \langle\ ,\ \rangle)$.

As in Section 5.3, let J denote the set of all signed indices j such that the eigenvalue λ^j of (6.1.1) of signed index j exists. This definition assumes that the eigenvalue problem (6.1.1) is locally definite. The set J and the eigenvalues λ^j are the same for (6.1.1) and (6.1.2). An *eigensystem* of (6.1.1) is a φ-orthogonal system $x^j \in E(\lambda^j)$, $j \in J$, of eigenvectors. A system x^j, $j \in J$, of tensors is an eigensystem of (6.1.1) if and only if it is of the form $x^j = Sy^j$, $j \in J$, where y^j, $j \in J$, is an eigensystem of (6.1.2).

6.4 Left definite eigenvalue problems

We consider the eigenvalue problem (6.1.1) which is assumed to be left definite with respect to $\mu_1,\dots,\mu_k$; see Section 3.4. Then the transformed problem (6.1.2) is also left definite, and Theorem 5.6.4 yields

THEOREM 6.4.1. *Let* x^j, $j \in J$, *be an eigensystem of the left definite problem (6.1.1). Then, for all* $x,y \in G$, $\varphi_0(x,y)$ *can be expanded into the absolutely convergent series*

$$\varphi_0(x,y) = \sum_{j\in J} \frac{\varphi_0(x,x^j)\,\varphi_0(x^j,y)}{\varphi_0(x^j,x^j)} \quad .$$

Proof. Define y^j, $j \in J$, by $x^j = Sy^j$. Then y^j, $j \in J$, is an eigensystem of (6.1.2). Hence the statement of our theorem follows from Theorem 5.6.4 and (6.3.3) noting that G is the range of S .□

The form

$$\psi = \sum_{s=1}^{k} \mu_s \varphi_s \tag{6.4.1}$$

defines an inner product in G . Let $(\overline{G}, \overline{\psi})$ be the completion of G with respect to this inner product. It follows from (6.3.3) that S is an isometry from $(H, \tilde{\psi})$ onto (G, ψ) , where $\tilde{\psi}$ is defined by (5.6.1) corresponding to problem (6.1.2). Let $\overline{S}$ be the isometry from $(\overline{H}, \overline{\tilde{\psi}})$ onto $(\overline{G}, \overline{\psi})$ induced by S . We define an operator Γ on the Hilbert space $(\overline{G}, \overline{\psi})$ by

$$\Gamma := \overline{S}\,\tilde{\Gamma}\,\overline{S}^{-1}$$

where $\tilde{\Gamma}$ is the operator of Section 5.7 corresponding to problem (6.1.2). Then Γ is the bounded linear operator on $(\overline{G}, \overline{\psi})$ which is uniquely determined by

$$\overline{\psi}(\Gamma x,y) = \varphi_0(x,y) \ , \quad x,y \in G \ . \tag{6.4.2}$$

Now Theorem 5.7.3 immediately gives

THEOREM 6.4.2. *The operator* Γ *is compact and Hermitian on* $(\overline{G}, \overline{\psi})$. *The nonzero eigenvalues of* Γ *are given by* $1/\mu\lambda$ *where* $\mu := (0,\mu_1,\ldots,\mu_k)$ *and* λ *runs through the set of eigenvalues of (6.1.1). The eigenspace of* Γ *associated with an eigenvalue* $1/\tau$ *is the direct sum of the eigenspaces of (6.1.1) associated with the finite number of eigenvalues* λ *satisfying* $\mu\lambda = \tau$.

We now assume that (6.1.1) is strictly left definite with respect to $\mu_1,\ldots,\mu_k$; see Section 3.4. We shall show that the above theorem can be improved under this assumption. Let ψ_ℓ denote the form determinant of the k by k array $\langle A_{rs}\cdot,\cdot\rangle_r$, $r,s = 1,\dots,k$, with ℓ^{th} row replaced by $\mu_1,\dots,\mu_k$. By Lemma 6.2.2, ψ_ℓ is continuous on $(\hat{H}_\ell, \langle\ ,\ \rangle)$, where

$$\hat{H}_\ell := H_1 \otimes \dots \otimes H_{\ell-1} \otimes H_{\ell+1} \otimes \dots \otimes H_k$$

and $\langle\ ,\ \rangle$ denotes the tensorial product of the inner products $\langle\ ,\ \rangle_r$, $r = 1,\dots,k,\ r \neq \ell$. By the assumed strict left definiteness, there is a positive ε such that

$$\psi_\ell(y,y) \geq \varepsilon \langle y,y\rangle \quad \text{for all decomposable} \quad y \in \hat{H}_\ell\ ,\ \ell = 1,\dots,k\ .$$

Hence, by Theorem 6.2.3, there are positive numbers η and θ such that

$$\theta \langle y,y\rangle \geq \psi_\ell(y,y) \geq \eta \langle y,y\rangle\ ,\quad y \in \hat{H}_\ell\ ,\ \ell = 1,\dots,k\ , \tag{6.4.3}$$

i. e. the inner products $\langle\ ,\ \rangle$ and ψ_ℓ are equivalent on $\hat{H}_\ell$.

Since $A_{\ell o}$ is negative definite for every $\ell = 1,\dots,k$, there is a positive δ such that

$$-\bar{\omega}_\ell(x_\ell,x_\ell) \geq \delta \langle x_\ell,x_\ell\rangle_\ell \quad \text{for } x_\ell \in G_\ell,\ \ell = 1,\dots,k\ . \tag{6.4.4}$$

If we expand the form determinant ψ of the array (4.5.3) with $\mu_o = 0$ according to the zeroth column then we obtain

$$\psi = \sum_{\ell=1}^{k} (-\bar{\omega}_\ell) \otimes \psi_\ell\ . \tag{6.4.5}$$

Now (6.4.3), (6.4.4), (6.4.5) and Lemma 6.2.2 yield the inequalities

$$k\,\theta\,\delta^{1-k}(-\bar{\omega}_1) \otimes \dots \otimes (-\bar{\omega}_k)(x,x) \geq \psi(x,x) \geq k\eta\delta \langle x,x\rangle \quad \text{for } x \in G\ .$$

Thus we have shown

LEMMA 6.4.3. *Under strict left definiteness, the form* $\langle\ ,\ \rangle$ *is continuous on* (G,ψ), *and the form* ψ *is continuous on* $(G, (-\bar{\omega}_1) \otimes \dots \otimes (-\bar{\omega}_k))$.

Of course, the second part of the lemma also holds under left definiteness. The above lemma shows that the canonical injection

$$\kappa : G \to H$$

is continuous from (G,ψ) into $(H, < , >)$. Hence κ admits a uniquely determined continuous extension

$$\bar{\kappa} : \bar{G} \to \bar{H} \quad , \tag{6.4.6}$$

where $(\bar{G}, \bar{\psi})$ and $(\bar{H}, < , >)$ are the completions of (G,ψ) and $(H, < , >)$, respectively.

LEMMA 6.4.4. *Under strict left definiteness, the map (6.4.6) is one-to-one. Hence* $\bar{G}$ *can be identified with a linear subspace of* $\bar{H}$.

Proof. We first note that, by definition of $\bar{\omega}_r$, D_r is dense in $(G_r, -\bar{\omega}_r)$. Hence $D = D_1 \otimes \ldots \otimes D_k$ is dense in G with respect to the inner product $(-\bar{\omega}_1) \otimes \ldots \otimes (-\bar{\omega}_k)$; see Lemma 6.2.1. It follows from Lemma 6.4.3 that D is dense in (G,ψ) . Consequently, D is dense in $(\bar{G}, \bar{\psi})$.

To prove that $\bar{\kappa}$ is one-to-one, let $x \in \bar{G}$ satisfy $\bar{\kappa}(x) = 0$. Then, by what we have just shown, there is a sequence $x^n \in D$ converging to x with respect to $\bar{\psi}$ and converging to 0 with respect to $< , >$. For each $n,m = 1,2,\ldots$, we write

$$\psi(x^n,x^n) \leq |\psi(x^n,x^n-x^m)| + |\psi(x^n,x^m)| \quad .$$

The first term on the right-hand side tends to 0 as $n,m \to \infty$. By Lemma 6.3.1 and (6.4.1), the second term tends to 0 as $m \to \infty$ for every fixed n . This shows that $\psi(x^n,x^n)$ tends to 0 as $n \to \infty$ which proves that $x = 0$. □

We mention that the statement of the above lemma just means that the form ψ is closable with respect to the Hilbert space $(\bar{H}, < , >)$.

We recall that the operator determinant Δ_0 is a continuous linear operator on $(H, < , >)$. Hence it admits a uniquely determined continuous extension $\bar{\Delta}_0$ on $(\bar{H}, < , >)$. By (6.3.5) and (6.4.2),

$$\bar{\psi}(\Gamma x,y) = <\bar{\Delta}_0 x,y> \quad \text{for all} \quad x,y \in G \quad .$$

The left-hand side of this equation depends continuously on $x,y \in \bar{G}$ with respect to $\bar{\psi}$. By Lemma 6.4.4, the right-hand side is well defined for $x,y \in \bar{G}$, and, by continuity of $\overline{\kappa}$, it also depends continuously on $x,y \in \overline{G}$ with respect to $\overline{\psi}$. Hence

$$\overline{\psi}(\Gamma x,y) = <\overline{\Delta}_0 x,y> \text{ for all } x,y \in \overline{G} . \qquad (6.4.7)$$

If $\Gamma x = 0$ for some $x \in \overline{G}$ then, by (6.4.7), $<\overline{\Delta}_0 x,y> = 0$ for all $y \in \overline{G}$. Since $\overline{G}$ is dense in $(\overline{H}, < , >)$, it follows that $\overline{\Delta}_0 x = 0$. Conversely, $\overline{\Delta}_0 x = 0$ for some $x \in \overline{G}$ implies that $\Gamma x = 0$. This proves

$$\text{Ker } \Gamma = (\text{Ker } \overline{\Delta}_0) \cap \overline{G} .$$

Hence Theorem 6.4.2 yields

THEOREM 6.4.5. *Every eigensystem* x^j, $j \in J$, *of the strictly left definite eigenvalue problem (6.1.1) satisfying* $\psi(x^j,x^j) = 1$, $j \in J$, *is an orthonormal basis of the orthocomplement of* $(\text{Ker } \overline{\Delta}_0) \cap \overline{G}$ *with respect to* $(\overline{G}, \overline{\psi})$.

6.5 Right definite eigenvalue problems

In this section we assume that the eigenvalue problem (6.1.1) is right definite i. e.

$$\varphi_0(x,x) > 0 \text{ for all nonzero decomposable } x \in G .$$

Then, by Theorem 5.2.1, φ_0 is an inner product in G. Since right definiteness of (6.1.1) implies right definiteness of the transformed problem (6.1.2), it is very simple to carry over the results of Section 5.8 to problem (6.1.1). In the following let x^j, $j \in J$, be an eigensystem of (6.1.1) satisfying $\varphi_0(x^j,x^j) = 1$ for each $j \in J$.

THEOREM 6.5.1. *If there are real numbers* $\mu_1,\ldots,\mu_k$ *such that the right definite problem (6.1.1) satisfies condition (2.7.2) for all unit vectors* $u_r \in G_r$ *then* x^j, $j \in J$, *is an orthonormal basis of* (G,φ_0).

Proof. Problem (6.1.2) satisfies the assumptions of Theorem 5.8.1. Hence $S^{-1} x^j$, $j \in J$, is an orthonormal basis of $(H,\tilde{\varphi}_0)$. Since S is an isometry from $(H,\tilde{\varphi}_0)$ onto (G,φ_0), this implies the statement of the theorem. □

THEOREM 6.5.2. *Let problem (6.1.1) be right definite, and let* $x \in G$ *be such that the linear functional* $\varphi_s(\cdot,x) = \overline{\varphi_s(x,\cdot)}$ *is continuous on* (G,φ_0) *for every* $s = 1,\dots,k$ *. Then there holds the Fourier expansion*

$$x = \sum_{j\in J} \varphi_0(x,x^j)x^j \quad .$$

Proof. The tensor $S^{-1}x$ satisfies the condition of Theorem 5.8.2 applied to (6.1.2) which guarantees that

$$S^{-1}x = \sum_{j\in J} \tilde{\varphi}_0(S^{-1}x\, ,\, S^{-1}x^j)S^{-1}x^j \quad .$$

This completes the proof because S is an isometry. □

Of course, also Theorem 5.8.3 can be carried over to problem (6.1.1). We only formulate a special case of the general result. We shall show that, under strict right definiteness of (6.1.1), the transformed problem (6.1.2) satisfies the assumptions of Theorem 5.8.3. So assume that there is a positive ε such that

$$\varphi_0(x,x) \geq \varepsilon \langle x,x\rangle \quad \text{for all decomposable} \quad x \in H \quad . \tag{6.5.1}$$

Since all operators A_{rs}, $r,s = 1,\dots,k$, are continuous, it follows that

$$\varphi_0'(x,x) \geq \frac{\varepsilon}{2}\langle x,x\rangle \quad \text{for all decomposable} \quad x \in H \quad , \tag{6.5.2}$$

whenever φ_0' is the form determinant of $\langle A_{rs}'\cdot\, ,\cdot\rangle_r$, $r,s = 1,\dots,k$, and A_{rs}' is a bounded Hermitian operator on H_r sufficiently close to A_{rs} with respect to the operator norm. In particular, there is a positive θ such that (6.5.2) holds for $A_{rs}' = A_{rs} + \theta_{rs} I_r$, $-\theta \leq \theta_{rs} \leq \theta$, $r,s = 1,\dots,k$. Hence the assumptions of Theorem 5.8.3 are satisfied for the transformed problem (6.1.2) if we choose $B_r = \theta S_r^2$, $r = 1,\dots,k$. It follows that, under strict right definiteness, the eigensystem x^j, $j \in J$, is an orthonormal basis of (G, φ_0) . We shall give a second proof of this important result based on Theorem (6.5.2). We first note

LEMMA 6.5.3. *If the eigenvalue problem (6.1.1) is strictly right definite then the inner products* φ_0 *and* $\langle\ ,\ \rangle$ *are equivalent on* H .

The proof follows from (6.5.1), Theorem 6.2.3 and from continuity of φ_0 with respect to $< , >$. □

THEOREM 6.5.4. *If the eigenvalue problem (6.1.1) is strictly right definite then* x^j, $j \in J$, *is an orthonormal basis of* (H, φ_0) .

Proof. Let $x \in D$. Then, by Lemma 6.3.1, the functionals $\varphi_s(\cdot,x)$, $s = 1,\dots,k$, are continuous on $(G, < , >)$. Hence, by Lemma 6.5.3, $\varphi_s(\cdot,x)$ is continuous on (G,φ_0) for every $s = 1,\dots,k$. Now Theorem 6.5.2 shows that x lies in the closure of the linear span of x^j, $j \in J$, with respect to φ_0 . Since D is dense in (H,φ_0) , this proves the theorem. □

6.6 Separation of variables

In the remaining sections of this chapter we shall specialize our results to the boundary eigenvalue problem (3.5.1), (3.5.2). Our general assumptions and notations are taken from Section 3.5. As in Section 5.9 , we identify the decomposable tensor $x = x_1 \otimes \dots \otimes x_k$, $x_r \in H_r = L^2(a_r,b_r)$, with the function

$$x(\xi) = x_1(\xi_1) \dots x_k(\xi_k),\ \xi = (\xi_1,\dots,\xi_k) \in \Pi = \prod_{r=1}^{k} [a_r,b_r] ,$$

which lies in $L^2(\Pi)$. Then $H = H_1 \otimes \dots \otimes H_k$ is a linear subspace of $L^2(\Pi)$, and the usual inner product $< , >$ in $L^2(\Pi)$ restricted to H is the tensorial product of the inner products $< , >_r$ in $L^2(a_r,b_r)$. Since H is dense in $L^2(\Pi)$, $(L^2(\Pi), < , >)$ is the completion of $(H, < , >)$.

The operator determinant Δ_0 is the multiplication operator

$$(\Delta_0 x)(\xi) = d_0(\xi)x(\xi) , \tag{6.6.1}$$

where

$$d_0(\xi) := \det_{1\le r,s\le k} a_{rs}(\xi_r),\ \xi = (\xi_1,\dots,\xi_k) \in \Pi . \tag{6.6.2}$$

The associated sesquilinear form φ_0 is given by

$$\varphi_0(x,y) = \int_{\Pi} d_0(\xi)x(\xi)\overline{y(\xi)}d\xi . \tag{6.6.3}$$

The operator determinants Δ_s, $s = 1,\dots,k$, are partial differential operators

$$-\Delta_s = \sum_{r=1}^{k} d_{rs}\left(\frac{\partial}{\partial\xi_r}\left(p_r\frac{\partial}{\partial\xi_r}\right)+q_r\right) \quad , \quad s = 1,\dots,k \quad , \tag{6.6.4}$$

where $(-1)^{\ell+t}\, d_{\ell t}(\xi_1,\dots,\xi_{\ell-1},\ \xi_{\ell+1},\dots,\xi_k)$ denotes the determinant of the matrix $a_{rs}(\xi_r)$, $r,s = 1,\dots,k$, with ℓ^{th} row and t^{th} column deleted. If $\lambda = (1,\lambda_1,\dots,\lambda_k)$ is an eigenvalue of (3.5.1), (3.5.2) and $x \in E(\lambda)$ is an associated eigenfunction then, by (6.3.4), (6.3.5), (6.3.6),

$$\Delta_s\, x = \lambda_s\, \Delta_o\, x, \quad 0 \neq x \in D, \quad s = 1,\dots,k \quad . \tag{6.6.5}$$

For any nonzero tuple $(\mu_1,\dots,\mu_k)$ of real numbers, (6.6.5) yields

$$\left(\sum_{s=1}^{k} \mu_s\Delta_s\right) x = \left(\sum_{s=1}^{k} \mu_s\lambda_s\right)\Delta_o\, x \ , \quad 0 \neq x \in D \quad . \tag{6.6.6}$$

The representations (6.6.1), (6.6.4) show that (6.6.6) is a boundary eigenvalue problem for a partial differential operator associated with the given multiparameter eigenvalue problem (3.5.1), (3.5.2). If we start with problem (6.6.6) then the eigenvalue problem (3.5.1), (3.5.2) is said to arise from (6.6.6) by the method of *separation of variables*. For instance, if $\mu_1 = \dots = \mu_{k-1} = 0$, $\mu_k = 1$, then λ_k is the eigenvalue parameter of problem (6.6.6) and $\lambda_1,\dots,\lambda_{k-1}$ are usually called *separation constants*. Examples for the method of separation of variables can be found in the introduction and in Section 6.9.

6.7 Left definite boundary eigenvalue problems

According to Theorem 3.6.2(iv), the boundary eigenvalue problem (3.5.1), (3.5.2) is left definite with respect to $\mu_1,\dots,\mu_k$ if and only if the differential operators A_{ro} are negative definite for each r , and, for every $\ell = 1,\dots,k$, the function

$$c_\ell(\hat{\xi}_\ell) = \sum_{s=1}^{k} \mu_s\, d_{\ell s}(\hat{\xi}_\ell) \quad , \quad \hat{\xi}_\ell \in \hat{\Pi}_\ell = \prod_{\substack{r=1\\ r\neq\ell}}^{k} [a_r,b_r] \quad , \tag{6.7.1}$$

is positive on a dense subset of $\hat{\Pi}_\ell$. Under this assumption, the results of Section 6.4 can be applied to the eigenvalue problem (3.5.1), (3.5.2).

We shall interpret these results in more detail if the eigenvalue problem is strictly left definite. Then, by Theorem 3.6.2(iii), the functions c_ℓ are positive on $\hat{\Pi}_\ell$ for each ℓ . Since c_ℓ is continuous on the compact set $\hat{\Pi}_\ell$, there are real numbers ε, θ such that

$$0 < \varepsilon \leq c_\ell(\hat{\xi}_\ell) \leq \theta \quad \text{for } \hat{\xi}_\ell \in \hat{\Pi}_\ell \ , \ \ell = 1,\dots,k \ . \tag{6.7.2}$$

This implies that the formal partial differential operator

$$\sum_{r=1}^{k} c_r \left(\frac{\partial}{\partial \xi_r} (p_r \frac{\partial}{\partial \xi_r}) + q_r \right)$$

is *uniformly elliptic* on Π i. e. there is a positive number δ such that

$$\sum_{r=1}^{k} c_r(\hat{\xi}_r)\, p_r(\xi_r)\, z_r^2 \;\geq\; \delta \sum_{r=1}^{k} z_r^2$$

for all $(z_1,\dots,z_k) \in \mathbb{R}^k$ and $\xi \in \Pi$. Hence, by (6.6.4), (6.7.1), the multiparameter eigenvalue problem (3.5.1), (3.5.2) arises from the elliptic boundary eigenvalue problem (6.6.6) by separation of variables. Here we recognize the close relationship between left definite multiparameter boundary eigenvalue problems (3.5.1), (3.5.2) and elliptic one-parameter boundary eigenvalue problems for partial differential operators. There is an extensive theory of elliptic boundary eigenvalue problems which could be used to obtain expansion theorems for strictly left definite eigenvalue problems (3.5.1), (3.5.2). We choose another approach based on the general results of Section 6.4.

In the following we shall work with the Sobolev space $W^1(\mathrm{int}(\Pi))$ where $\mathrm{int}(\Pi)$ denotes the interior of Π ; see [Yoshida(1971), Chapter I, Section 9]. For abbreviation, we denote this space by $W^1(\Pi)$. The space $W^1(\Pi)$ consists of those functions x on $\mathrm{int}(\Pi)$ which together with their first derivatives $\partial x / \partial \xi_r$, $r = 1,\dots,k$, lie in $L^2(\Pi)$. The derivatives have to be understood in the sense of distributions. The space $W^1(\Pi)$ is a Hilbert space under the inner product

$$\langle x,y \rangle + \sum_{r=1}^{k} \langle \frac{\partial x}{\partial \xi_r} , \frac{\partial y}{\partial \xi_r} \rangle \ , \tag{6.7.3}$$

$\langle \ , \ \rangle$ denoting the inner product in $L^2(\Pi)$. We remark that G is a linear subspace of the Sobolev space $W^1(\Pi)$ because

$$G = G_1 \otimes \dots \otimes G_k \subset W^1(a_1,b_1) \otimes \dots \otimes W^1(a_k,b_k) \subset W^1(\Pi) \ ;$$

see Lemma 3.2.3.

We now show that the inner product (6.4.1) is equivalent to the inner product (6.7.3) of the Sobolev space $W^1(\Pi)$. This statement is closely related to Gårding's inequality [Yoshida(1971), Chapter IV, Section 8] valid for uniformly elliptic partial differential operators.

LEMMA 6.7.1. *If the eigenvalue problem (3.5.1), (3.5.2) is strictly left definite then the inner product (6.7.3) is equivalent to* ψ *on* G .

Proof. By (6.4.5), we can write ψ as

$$\psi = \sum_{\ell=1}^{k} (-\bar{\omega}_\ell) \otimes \psi_\ell . \tag{6.7.4}$$

The form ψ_ℓ is given by

$$\psi_\ell(x,y) = \int_{\hat{\Pi}_\ell} c_\ell(\hat{\xi}_\ell)\, x(\hat{\xi}_\ell)\, \overline{y(\hat{\xi}_\ell)}\, d\hat{\xi}_\ell .$$

Hence (6.7.2) shows that the inner product ψ_ℓ is equivalent to the usual inner product of $L^2(\hat{\Pi}_\ell)$.

By (6.4.4), the forms $-\bar{\omega}_r$ are inner products in G_r . It follows from the inequalities (3.2.9), (3.2.10) that $-\bar{\omega}_r$ is equivalent to the usual inner product of the Sobolev space $W^1(a_r,b_r)$.

If we replace the inner products $-\bar{\omega}_\ell$ and ψ_ℓ appearing in the representation (6.7.4) by their equivalent inner products then, by Lemma 6.2.2, we obtain an inner product which is equivalent to ψ . This inner product is given by

$$\sum_{r=1}^{k} \left(\langle x,y\rangle + \langle \frac{\partial x}{\partial \xi_r} , \frac{\partial y}{\partial \xi_r} \rangle \right) , \quad x,y \in G ,$$

which, obviously, is equivalent to the inner product (6.7.3). This completes the proof.□

In Section 6.4 we used the completion $(\bar{G}, \bar{\psi})$ of the space (G, ψ) . By virtue of the above lemma, $\bar{G}$ is the closure of G with respect to the Sobolev space $W^1(\Pi)$, and the inner product $\bar{\psi}$ is equivalent to the inner product (6.7.3). Hence it follows from (3.2.12) that

$$W_0^1(\Pi) \subset \overline{G} \subset W^1(\Pi) \quad ,$$

where $W_0^1(\Pi)$ denotes the closure of $C_0^\infty(\mathrm{int}(\Pi))$ with respect to $W^1(\Pi)$. We used the fact that $W_0^1(a_1,b_1) \otimes \ldots \otimes W_0^1(a_k,b_k)$ is dense in $W_0^1(\Pi)$; see [Yoshida (1971), Chapter 1, Section 14, Theorem 1]. Similarly, $W^1(a_1,b_1) \otimes \ldots \otimes W^1(a_k,b_k)$ is dense in $W^1(\Pi)$. For example, we have $\overline{G} = W_0^1(\Pi)$ under the boundary conditions

$$x_r(a_r) = x_r(b_r) = 0, \; r = 1,\ldots,k \quad ,$$

and $\overline{G} = W^1(\Pi)$ under the boundary conditions

$$x_r'(a_r) + \alpha_r \, x_r(a_r) = 0, \quad x_r'(b_r) + \beta_r \, x_r(b_r) = 0, \quad r = 1,\ldots,k \quad .$$

Now Lemma 6.7.1 and Theorem 6.4.5 yield

THEOREM 6.7.2. *Let the boundary eigenvalue problem (3.5.1), (3.5.2) be strictly left definite, and assume that the set of zeros of the function* d_0 *has Lebesgue measure zero. Then every eigensystem* x^j, $j \in J$, *of (3.5.1), (3.5.2) with* $\psi(x^j,x^j) = 1$, $j \in J$, *is an orthonormal basis of the Hilbert space* $(\overline{G}, \overline{\psi})$. *For every function* $x \in \overline{G}$, *the Fourier expansion*

$$x = \sum_{j \in J} \overline{\psi}(x,x^j) \, x^j$$

converges with respect to the usual inner product of the Sobolev space $W^1(\Pi)$.

Proof. By Lemma 6.7.1 and Theorem 6.4.5, it suffices to prove that $\overline{G} \cap \mathrm{Ker}\, \overline{\Delta}_0$ contains only the zero function. The operator $\overline{\Delta}_0$ is the continuous extension of the multiplication operator (6.6.1) onto $L^2(\Pi)$. Hence

$$(\overline{\Delta}_0 x)(\xi) = d_0(\xi) \, x(\xi) \quad , \quad x \in L^2(\Pi) \quad .$$

Since the set of zeros of d_0 has measure zero, $\overline{\Delta}_0$ has trivial kernel which completes the proof.□

6.8 Right definite boundary eigenvalue problems

Let the eigenvalue problem (3.5.1), (3.5.2) be strictly right definite. By Lemma

3.6.2(i) this means that

$$d_0(\xi) > 0 \quad \text{for all} \quad \xi \in \Pi \ . \tag{6.8.1}$$

Since d_0 is continuous on the compact set Π, there are real numbers ε, θ such that

$$0 < \varepsilon \le d_0(\xi) \le \theta \quad \text{for all} \quad \xi \in \Pi \ .$$

Hence the inner product φ_0 is equivalent to the usual inner product in $L^2(\Pi)$. Here we consider φ_0 as a form on $L^2(\Pi)$ defined by (6.6.3). Since G is dense in $L^2(\Pi)$, it follows that $(L^2(\Pi), \varphi_0)$ is the completion of (G, φ_0). Therefore Theorem 6.5.4 yields

THEOREM 6.8.1. *If the boundary eigenvalue problem (3.5.1), (3.5.2) satisfies (6.8.1) then every eigensystem* x^j, $j \in J$, *of (3.5.1), (3.5.2) with* $\varphi_0(x^j, x^j) = 1$, $j \in J$, *is an orthonormal basis of the Hilbert space* $(L^2(\Pi), \varphi_0)$.

Our aim is now to prove completeness of eigensystems under the weaker assumption of right definiteness. By Lemma 3.6.2 (ii), right definiteness means that the open set

$$\Omega := \{\xi \in \mathrm{int}(\Pi) \mid d_0(\xi) > 0\} \text{ is dense in } \Pi \ . \tag{6.8.2}$$

Since d_0 is continuous, it follows that the form φ_0 is positive semidefinite on $L^2(\Pi)$ and positive definite on the space $C(\Pi)$ of continuous functions on Π. We remark that, in general, φ_0 is not positive definite on $L^2(\Pi)$ because the set of zeros of d_0 may have positive Lebesgue measure.

We need the following approximation lemma.

LEMMA 6.8.2. *The set of functions*

$$x = x_1 \otimes \ldots \otimes x_k \in C_0^\infty(\Omega), \quad x_r \in C_0^\infty(a_r, b_r) \quad , \tag{6.8.3}$$

i. e. the set of products of functions $x_r \in C_0^\infty(a_r, b_r)$ *which vanish outside a compact subset of* Ω, *is total in* $(L^2(\Pi), \varphi_0)$.

Proof. Let $\Omega_1 \times \ldots \times \Omega_k$ be an open rectangle contained in Ω. The characteristic function of Ω_r is the L^2-limit of a sequence $x_r^n \in C_0^\infty(\Omega_r)$. Hence the characteristic function of $\Omega_1 \times \ldots \times \Omega_k$ is the L^2-limit of the sequence $x_1^n \otimes \ldots \otimes x_k^n$. This shows that all characteristic functions of open rectangles contained in Ω are L^2-limits of sequences of functions of the form (6.8.3). Since Ω is open, it is well known that these characteristic functions form a total subset of $L^2(\Omega)$. Hence the set (6.8.3) is total in $L^2(\Omega)$ and, since L^2-convergence implies φ_0-convergence, is total in $(L^2(\Omega), \varphi_0)$, too. Now, by definition of Ω, it is trivial that $L^2(\Omega)$ is dense in $(L^2(\Pi), \varphi_0)$ which completes the proof. □

THEOREM 6.8.3. *Under assumption (6.8.2), every eigensystem* x^j, $j \in J$, *of the boundary eigenvalue problem (3.5.1), (3.5.2) satisfying* $\varphi_0(x^j, x^j) = 1$, $j \in J$, *is an orthonormal basis of the semidefinite inner product space* $(L^2(\Pi), \varphi_0)$.

Proof. By Lemma 6.8.2, it will be sufficient to prove that every function of the form (6.8.3) can be expanded into a Fourier series of the given eigensystem. So let $x = x_1 \otimes \ldots \otimes x_k$, $x_r \in C_0^\infty(a_r, b_r)$, vanish outside a compact subset K of Ω. Then the representation (6.6.4) of the partial differential operator Δ_s immediately shows that $\Delta_s x$ is a continuous function on Π vanishing outside K for every $s = 1, \ldots, k$. Since the continuous function d_0 is positive on the compact set K, there is a positive ε such that $d_0(\xi) \geq \varepsilon$ for $\xi \in K$. Hence, by (6.3.6),

$$\begin{aligned}
|\varphi_s(x,y)|^2 &= |\langle \Delta_s x, y\rangle|^2 \\
&= \left| \int_K (\Delta_s x)(\xi)\, \overline{y(\xi)}\, d\xi \right|^2 \leq \langle \Delta_s x, \Delta_s x\rangle \int_K |y(\xi)|^2\, d\xi \\
&\leq \frac{1}{\varepsilon} \langle \Delta_s x, \Delta_s x\rangle \int_K d_0(\xi)|y(\xi)|^2\, d\xi \\
&\leq \frac{1}{\varepsilon} \langle \Delta_s x, \Delta_s x\rangle\, \varphi_0(y,y) \qquad \text{for all } y \in G .
\end{aligned}$$

This proves continuity of the functionals $\varphi_s(x,.)$, $s = 1, \ldots, k$, on (G, φ_0). Hence, by Theorem 6.5.2, x can be expanded into a Fourier series of the given eigensystem. □

6.9 Separation in ellipsoidal coordinates

Let $e_1,\dots,e_k$ be fixed real numbers which satisfy

$$e_1 < e_2 < \dots < e_k < e_{k+1} := \infty .$$

The ellipsoidal coordinates $\xi = (\xi_1,\dots,\xi_k)$ of a point in $]0,\infty[^k$ are related to its cartesian coordinates $\eta = (\eta_1,\dots,\eta_k)$ by

$$\sum_{r=1}^{k} \frac{\eta_r^2}{\xi_\ell - e_r} = 1 , \quad \xi_\ell \in]e_\ell,e_{\ell+1}[, \quad \ell = 1,\dots,k .$$

Thereby the domain

$$\eta_r \in]0,\infty[, \quad r = 1,\dots,k ,$$

is mapped analytically one-to-one onto the domain

$$\xi_r \in]e_r,e_{r+1}[, \quad r = 1,\dots,k .$$

The inverse mapping is given by

$$\eta_r^2 = (\xi_r - e_r) \prod_{\substack{\ell=1 \\ \ell \neq r}}^{k} \frac{\xi_\ell - e_r}{e_\ell - e_r} , \quad r = 1,\dots,k .$$

We refer to [Meixner and Schäfke (1954), Section 1.121] where the three dimensional case is treated; the general case is analogous.

If we transform the k - dimensional wave equation

$$\sum_{r=1}^{k} \frac{\partial^2 y}{\partial \eta_r^2} + \nu y = 0 \tag{6.9.1}$$

into ellipsoidal coordinates then we obtain, after separation of variables

$$y(\eta) = x(\xi) = x_1(\xi_1)\dots x_k(\xi_k) ,$$

the ordinary differential equations

$$(px_r')' + \frac{(-1)^{k-r}}{4p(\xi_r)} \left(\sum_{s=1}^{k} \lambda_s \xi_r^{s-1} \right) x_r = 0 , \quad r = 1,\dots,k , \tag{6.9.2}$$

where

$$p(t) := \left(\prod_{\ell=1}^{k} |t - e_\ell| \right)^{1/2} .$$

Here $\lambda_k = \nu$ is the eigenvalue parameter of (6.9.1) and $\lambda_1,\dots,\lambda_{k-1}$ are the separation constants; see [Meixner and Schäfke (1954), Section 1.131] for the case $k = 3$ and [Schmidt and Wolf (1979), Section 1.2] for the general case.

We now pose the problem to solve the wave equation (6.9.1) subject to the condition that the function y vanishes on the boundary of the domain V, where V is given by

$$\xi_r \in [a_r,b_r] \subset]e_r,e_{r+1}[, \quad r = 1,\dots,k .$$

If $k = 2$ then V looks like this

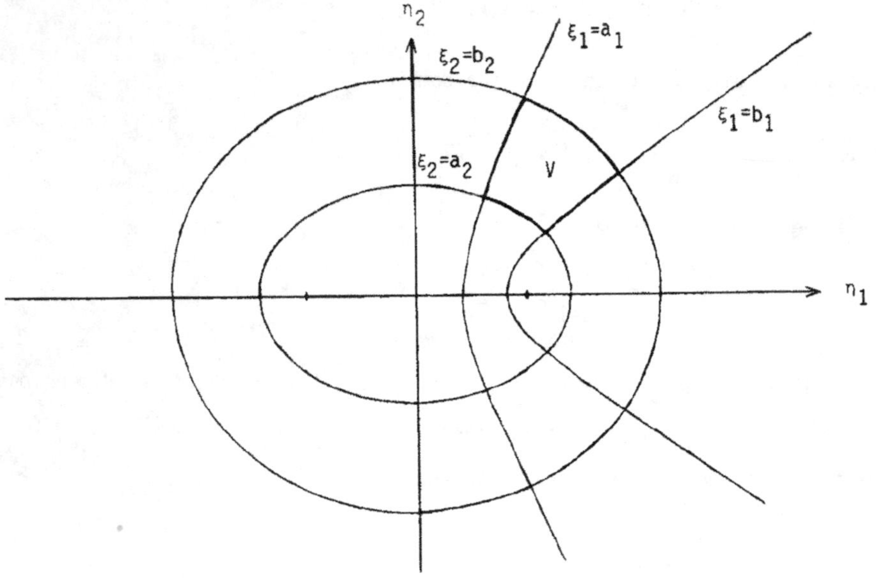

We are thus led to solve the differential equations (6.9.2) subject to the boundary conditions

$$x_r(a_r) = x_r(b_r) = 0 , \quad r = 1,\dots,k . \tag{6.9.3}$$

This is a boundary eigenvalue problem of the form (3.5.1), (3.5.2).

Using the Vandermonde determinant, we can write d_0 as

$$d_0(\xi) = (-1)^{k(k-1)/2} \left(\prod_{r=1}^{k} 4p(\xi_r) \right)^{-1} \det \begin{pmatrix} 1 & \xi_1 & \cdots & \xi_1^{k-1} \\ \vdots & \vdots & & \vdots \\ 1 & \xi_k & \cdots & \xi_k^{k-1} \end{pmatrix}$$

$$= (-1)^{k(k-1)/2} \left(\prod_{r=1}^{k} 4p(\xi_r) \right)^{-1} \prod_{1\le r<s\le k} (\xi_s - \xi_r) \quad .$$

This shows that

$$(-1)^{k(k-1)/2}\, d_0(\xi) > 0 \quad \text{for} \quad \xi \in \Pi = \prod_{r=1}^{k} [a_r,b_r] \quad .$$

Similarly, we see that the eigenvalue problem (6.9.2), (6.9.3) is strictly left definite with respect to $\mu_1 = \ldots = \mu_{k-1} = 0$, $\mu_k = (-1)^{k(k-1)/2}$. Hence the Theorems 3.5.1, 3.5.2, 6.7.2 yield

THEOREM 6.9.1. *For given nonnegative integers* $n_1,\ldots,n_k$, *there are uniquely determined real numbers* $\lambda_1,\ldots,\lambda_k$ *such that the equations (6.9.2) admit real-valued solutions* x_r *on* $[a_r,b_r]$ *satisfying the boundary conditions (6.9.3) and having exactly* n_r *zeros in* $]a_r,b_r[$ *for every* $r = 1,\ldots,k$. *These functions* x_r *are uniquely determined up to a nonzero factor. If we form the products* $x(\xi) = x_1(\xi_1)\ldots x_k(\xi_k)$ *for every tuple* $(n_1,\ldots,n_k)$ *then, after normalization, we obtain a system of functions on* Π *which is an orthonormal basis of* $(W_0^1(\Pi), \overline{\psi})$. *The corresponding Fourier series of a function in* $W_0^1(\Pi)$ *converges with respect to the usual inner product (6.7.3) of* $W_0^1(\Pi)$.

In particular, the above completeness result shows that ν is an eigenvalue of (6.9.1) subject to the condition that $y(\eta) = 0$ on the boundary of V if and only if there is an eigenvalue $(\lambda_1,\ldots,\lambda_k)$ of (6.9.2), (6.9.3) satisfying $\lambda_k = \nu$. We further remark that the compact Hermitian operator $\pm\Gamma$ on $(W_0^1(\Pi), \overline{\psi})$ associated with (6.9.2), (6.9.3) according to Section 6.4 is the inverse of the Laplace operator transformed into ellipsoidal coordinates.

Of course, the results of this section can be generalized and modified in various ways. The wave equation (6.9.1) can be replaced by the generalized Schrödinger equation

$$\sum_{r=1}^{k} \frac{\partial^2 y}{\partial \eta_r^2} + \sum_{r=1}^{k} v_r(\eta) \frac{\partial y}{\partial \eta_r} + (w(\eta) + \nu) y = 0 \quad . \tag{6.9.4}$$

If the coefficients v_r and w are suitably chosen then (6.9.4) separates in ellipsoidal coordinates; see Schmidt and Wolf (1979). The general ellipsoidal coordinates can be replaced by other coordinate systems which arise from ellipsoidal coordinates by suitable limiting processes. We can also consider boundary conditions which are more general than (6.9.3).

6.10 The expansion in series of Lamé products

Let us consider the two-parameter eigenvalue problem (3.8.3), (3.8.4). Its eigenfunctions are of the form $E(\mathrm{sn}^2\xi_1)\, E(\mathrm{sn}^2(K+i\xi_2))$, $0 \le \xi_1 \le K$, $0 \le \xi_2 \le K'$, where E is a Lamé polynomial. We apply Theorem 6.8.3 (or Theorem 6.5.1) to this problem, and then we transform the completeness result back to Lamé's equation (3.7.1) where $e_1 = 0$, $e_2 = 1$, $e_3 = k^{-2}$. Omitting the details we only state the result which we obtain in this way. For $n = 0,2,4,\dots$ we denote the Lamé polynomials of degree $n/2$ by E_n^m, $m = 0,\dots,n/2$, according to Stieltjes' Theorem 3.8.3. We normalize the Lamé polynomials so that the Lamé products

$$(E_n^m \otimes E_n^m)(\zeta_1,\zeta_2) = E_n^m(\zeta_1)E_n^m(\zeta_2) \quad , n = 0,2,4,\dots, \ m = 0,\dots,n/2 \ , \tag{6.10.1}$$

satisfy

$$\langle E_n^m \otimes E_n^m , E_n^m \otimes E_n^m \rangle_R = 1 \quad , \tag{6.10.2}$$

where the inner product $\langle\ ,\ \rangle_R$ is defined by

$$\langle g_1, g_2\rangle_R = \frac{1}{2\pi} \int_0^1 \int_1^{k^{-2}} \frac{(\zeta_2 - \zeta_1) g_1(\zeta_1,\zeta_2)\overline{g_2(\zeta_1,\zeta_2)}}{w(\zeta_1)w(\zeta_2)}\, d\zeta_2 d\zeta_1 \quad , \tag{6.10.3}$$

and

$$w(\alpha) = (|\alpha||\alpha - 1||\alpha - k^{-2}|)^{1/2} \quad .$$

Then we have the following

THEOREM 6.10.1. *The Lamé products (6.10.1) normalized according to (6.10.2) form a system of functions on* $R =]0,1[\times]1,k^{-2}[$ *which is orthonormal and complete with respect to the inner product* $< , >_R$. *Every function* g *square-integrable with respect to* $< , >_R$ *can be expanded in the* $< , >_R$ *- convergent series*

$$g(\zeta_1,\zeta_2) \sim \sum_{\substack{n=0 \\ n \text{ even}}}^{\infty} \sum_{m=0}^{n/2} <g,E_n^m \otimes E_n^m>_R \; E_n^m(\zeta_1)E_n^m(\zeta_2) \quad , \quad (\zeta_1,\zeta_2) \in R \ . \qquad (6.10.4)$$

This theorem admits a simple interpretation in the theory of special functions by means of the sphero-conal coordinates in the two-dimensional unit sphere

$$S = \{(\eta_1,\eta_2,\eta_3) \in \mathbb{R}^3 \mid \eta_1^2 + \eta_2^2 + \eta_3^2 = 1\} \ .$$

The sphero-conal coordinates $0 \le \zeta_1 \le 1 \le \zeta_2 \le k^{-2}$, of a point (η_1,η_2,η_3) in S are defined as the solutions of the quadratic equation in α

$$\frac{\eta_1^2}{\alpha} + \frac{\eta_2^2}{\alpha-1} + \frac{\eta_3^2}{\alpha-k^{-2}} = 0 \quad , \quad \alpha = \zeta_1,\zeta_2 \ . \qquad (6.10.5)$$

If g_1,g_2 are two functions on R and $f_j(\eta_1,\eta_2,\eta_3) = g_j(\zeta_1,\zeta_2)$ are the corresponding functions on S then these functions are even with respect to each variable η_1,η_2,η_3 and the inner product (6.10.3) becomes

$$<g_1,g_2>_R = <f_1,f_2>_S := \frac{1}{4\pi} \int_S f_1(\eta)\overline{f_2(\eta)}d\eta$$

which is the usual inner product in the space $L^2(S)$ of square-integrable functions on the unit sphere S . In particular, if we transform the Lamé products (6.10.1) to functions on S by means of the sphero-conal coordinates (6.10.5) then we obtain spherical surface harmonics of degree n which are even (with respect to each variable η_1,η_2,η_3); see [Erdelyi, Magnus, Oberhettinger, Tricomi (1955), Section 15.7]. A spherical surface harmonic of degree n is a function defined on the unit sphere S which is the restriction of a harmonic polynomial homogeneous in η_1,η_2,η_3 of degree n onto S . For each fixed n , the $\frac{n}{2} + 1$ functions on S which we obtain from the Lamé products form a basis of the linear space of spherical surface harmonics of degree n which are even. Hence Theorem 6.10.1 just says that the spherical surface harmonics which are even are complete in the subspace of $L^2(S)$ consisting of even functions. If we consider also the other seven types of Lamé functions then we obtain the completeness of the spherical surface harmonics in $L^2(S)$. Of course, this result is well known; see [Sansone (1959), Chapter III, Section 20] . The expansion (6.10.4) in series of Lamé products then turns out to be the ordinary Laplace series in terms of sphero-conal coordinates.

6.11 Notes for Chapter 6

Theorem 6.2.3 was proved by Binding (1980a).

The expansion Theorem 6.4.5 for the strictly left definite eigenvalue problem (6.1.1) was given by Binding (1982c), (1982d). The expansion Theorems 6.4.1, 6.4.2 for left definite problems are new; compare Binding, Källström and Sleeman (1982).

Theorem 6.5.1 was proved by Cordes (1954) in the special case $k = 2$. Cordes assumes that the operators $A_{11}, A_{12}, -A_{21}, A_{22}$ are positive definite. It follows from this condition that the assumptions of Theorem 6.5.1 are satisfied with $\mu_1 = 0$ and $\mu_2 = 1$.

Theorem 6.5.2 is similar to results of [Atkinson (1972), Theorem 11.8.1], Browne (1977b), [Binding, Källström and Sleeman (1982), Theorem 5.1]; compare also Volkmer (1982a), (1982b). The completeness Theorem 6.5.4 for strictly right definite problems was shown by Källström and Sleeman (1976) and Browne (1977a).

The method of separation of variables was studied by many authors; see for instance Meixner and Schäfke (1954), Schmidt and Wolf (1979), Arscott and Darai (1981).

Expansion theorems for strictly left definite Sturm-Liouville problems with two parameters were investigated by Dixon (1907) and [Hilbert (1912), Chapter 21]. We refer also to Sleeman (1973a) and Faierman (1974), (1981a), (1982a), (1983). The general k - parameter case was treated by Källström and Sleeman (1974/75). Several of these papers use results for uniformly elliptic differential operators in their proofs of the expansion theorem. Hence their approach is different from ours.

The expansion Theorem 6.8.1 for strictly right definite boundary eigenvalue problems was proved by Faierman (1969) and Browne (1972). The completeness Theorem 6.8.3 in the right definite case seems to be new.

If we have established completeness of the eigenfunctions of a multiparameter Sturm-Liouville system then the next question is to ask for conditions on the function to be expanded which guarantee uniform convergence of the expansion. This type of problem has been intensively studied in various papers of Faierman and Roach.

Expansions in series of Lamé products have been investigated by several authors, for instance, by Dixon (1902) and Hilb (1907b). There are also expansions of holomorphic functions of two variables in series of Lamé products, see Volkmer (1988a,b). Theorem 6.10.1 is the starting point for that theory but then we have to use special methods which do not belong to multiparameter spectral theory.

BIBLIOGRAPHY

This bibliography contains quite a number of references in addition to those cited in the text.

Allahverdiev, B.P. and Isaev, H.A. (1981). Oscillation theorems for multiparameter problems with boundary conditions depending on spectral parameters, Izvestija Akad.Nauk.Azerb. SSR (6), 17-23.

Almamedov, M.S. and Isaev, H.A. (1985). Solvability of nonselfadjoint linear operator systems and the set of decomposability of multiparameter spectral problems, Soviet Math.Dokl.31, 472-474.

Almamedov, M.S.; Aslanov, A.A. and Isaev, H.A. (1985). On the theory of two-parameter spectral problems, Soviet Math.Dokl.32, 225-227.

Arscott, F.M. (1964a). Two-parameter eigenvalue problems in differential equations, Proc.London Math.Soc.14, 459-470.

Arscott, F.M. (1964b). Periodic Differential Equations, Pergamon Press, London.

Arscott, F.M. (1974a). Transform theorems for two-parameter eigenvalue problems in Hilbert space, Proc. Conference on Ordinary and Partial Differential Equations Dundee 1974, Lecture Notes in Mathematics 415, 302-307, Springer, Berlin.

Arscott, F.M. (1974b). Integral equation formulation of two-parameter eigenvalue problems, Conference on Spectral Theory and Asymptotics of Differential Equations, Scheveningen 1973, North Holland Math.Studies 13, 95-102.

Arscott, F.M. and Darai, A. (1981). Curvilinear coordinate systems in which the Helmholtz equation separates, IMA J.Appl.Math.27, 33-70.

Atkinson, F.V. (1951). A spectral problem for completely continuous operators, Acta.Math.Acad.Sci.Hungaricae 3, 53-60.

Atkinson, F.V. (1963). Boundary value problems leading to orthogonal polynomials in several variables, Bull.Amer.Soc.69, 345-351.

Atkinson, F.V. (1964a). Multivariate spectral theory: the linked eigenvalue problem for matrices. Technical Summary report No.431, Math. Research Center (U.S. Army), Madison, Wisconsin.

Atkinson, F.V. (1964b). Discrete and continuous boundary value problems, Academic Press, New York.

Atkinson, F.V. (1968). Multiparameter spectral theory, Bull.Amer.Math.Soc.74, 1-27.

Atkinson, F.V. (1972). Multiparameter Eigenvalue Problems, Vol.I: Matrices and Compact Operators, Academic Press, New York.

Atkinson, F.V. (1977). Deficiency-index theory in the multi-parameter Sturm-Liouville case, Proc.International Conference on Differential Equations, Uppsala, Sympos.Univ.Upsaliensis Ann.Quingent.Celebr. No.7, Almquist and Wiksell, Stockholm, 1-40.

Bailey, P.B. (1981). The automatic solution of two parameter Sturm-Liouville eigenvalue problems in ordinary differential equations, App.Math.and Comp.8, 251-259.

Berman, A. and Plemmons, R.J. (1979). Nonnegative matrices in the mathematical sciences, Academic Press, New York.

Binding, P. (1980a). Another positivity result for determinantal operators, Proc.Roy.Soc.Edinburgh 86A, 333-337.

Binding, P. (1980b). On the use of degree theory for nonlinear multiparameter eigenvalue problems, J.Math.Anal.Appl.73, 381-391.

Binding, P. (1981a). Multiparameter definiteness conditions, Proc.Roy.Soc. Edinburgh 89A, 319-332.

Binding, P. (1981b). Variatonal methods for one and several parameter non-linear eigenvalue problems, Can.J.Math.33, 210-228.

Binding, P. (1981c). On generalised and quadratic eigenvalue problems, Applicable Analysis 12, 27-45.

Binding, P. (1982a). Multiparameter definiteness conditions II, Proc.Roy.Soc. Edinburgh 93A, 47-61. Erratum: ibid 103A, 359.

Binding, P. (1982b). On a problem of B.D.Sleeman, J.Math.Anal.Appl.85, 291-307. Erratum: ibid 90, 270-271.

Binding, P. (1982c). Left definite multiparameter eigenvalue problems, Trans. Amer.Math.Soc.272, 475-486.

Binding, P. (1982d). Multiparameter variational principles, SIAM J.Math.Anal. 13, 842-855.

Binding, P. (1983a). Dual variational approaches to multiparameter eigenvalue problems, J.Math.Anal.Appl.92, 96-113.

Binding, P. (1983b). Abstract oscillation theorems for multiparameter eigen-value problems, J.Differential Equations 49, 331-343.

Binding, P. (1984a). The inertia of a Hermitian pencil, Lin.Alg.Appl.63, 179-191.

Binding, P. (1984b). Perturbation and bifurcation of nonsingular multi-parametric eigenvalues, Nonlinear Analysis 8, 335-352.

Binding, P. (1984c). Nonuniform right definiteness, J.Math.Anal.Appl.102, 233-243.

Binding, P. (1984d). Indicial equivalents of multiparameter definiteness conditions in finite dimensions, Proc.Edinburgh Math.Soc. 27, 283-296.

Binding, P. and Browne, P.J. (1977). A variational approach to multiparameter eigenvalue problems for matrices, SIAM J.Math.Anal.8, 763-777.

Binding, P. and Browne, P.J. (1978a). A variational approach to multiparameter eigenvalue problems in Hilbert space, SIAM J.Math.Anal.9, 1054-1067.

Binding, P. and Browne, P.J. (1978b). Positivity results for determinantal operators, Proc.Roy.Soc.Edinburgh 81A, 267-271.

Binding, P. and Browne, P.J. (1980). Comparison cones for multiparameter eigenvalue problems, J.Math.Anal.Appl.77, 132-149.

Binding, P. and Browne, P.J. (1981). Spectral properties of two-parameter eigenvalue problems, Proc.Roy.Soc.Edinburgh 89A, 157-173.

Binding, P. and Browne, P.J. (1983). A definiteness result for determinantal operators, ordinary differential equations and operators, Lecture Notes in Math.1032, 17-30, Springer, Berlin.

Binding, P. and Browne, P.J. (1984). Multiparameter Sturm theory, Proc.Roy.Soc. Edinburgh 99A, 173-184.

Binding, P. and Browne, P.J. (1987). Spectral properties of two parameter eigenvalue problems II, Proc.Roy.Soc.Edinburgh 106A, 39-51.

Binding, P. and Browne, P.J. (1988). Applications of two parameter spectral theory to symmetric generalised eigenvalue problems, Applicable Anal.29, 107-142.

Binding, P.; Browne, P.J. and Turyn, L. (1981). Existence conditions for two-parameter eigenvalue problems, Proc.Roy.Soc.Edinburgh 91A, 15-30.

Binding, P.; Browne, P.J. and Turyn, L. (1984a). Existence conditions for eigenvalue problems generated by compact multiparameter operators, Proc. Roy.Soc.Edinburgh 96A, 261-274.

Binding, P.; Browne, P.J. and Turyn, L. (1984b). Spectral properties of compact multiparameter operators, Proc.Roy.Soc.Edinburgh 98A, 291-303.

Binding, P.; Browne, P.J. and Turyn, L. (1986). Existence conditions for higher order eigensets of multiparameter operators, Proc.Roy.Soc. Edinburgh 103A, 137-146.

Binding, P.; Källström, A and Sleeman B.D. (1982). An abstract multiparameter spectral theory, Proc.Roy.Soc.Edinburgh 92A, 193-204.

Binding, P. and Seddighi, K. (1987a). Elliptic multiparameter eigenvalue problems, Proc.Edinburgh Mat.Soc. 30, 215-228.

Binding, P. and Seddighi, K. (1987b). On root vectors of self-adjoint pencils, J.Funct.Anal.70, 117-125.

Binding, P. and Volkmer, H. (1986). Existence and uniqueness of indexed multiparametric eigenvalues, J.Math.Anal.Appl.116, 131-146.

Bôcher, M. (1897/98a). The theorems of oscillation of Sturm and Klein I, Bull.Amer.Math.Soc.4, 295-313.

Bôcher, M. (1897/98b). The theorems of oscillation of Sturm and Klein II, Bull.Amer.Math.Soc.4, 365-376.

Bôcher, M. (1897/98c). The theorems of oscillation of Sturm and Klein III, Bull.Amer.Math.Soc.5, 22-43.

Bohte, Z. (1968). On solvability of some two-parameter eigenvalue problems in Hilbert space, Proc.Roy.Soc.Edinburgh 68A, 83-93.

Browne, P.J. (1972a). A multi-parameter eigenvalue problem, J.Math.Anal.Appl. 38, 553-568.

Browne, P.J. (1972b). A singular multi-parameter eigenvalue problem in second order ordinary differential equations, J.Differential Equations 12, 81-94.

Browne, P.J. (1974a). Multi-parameter spectral theory, Indiana Univ.Math.J. 24, 249-257.

Browne, P.J. (1974b). Multi-parameter problems, Proc. Conference on Ordinary and Partial Differential Equations Dundee 1974, Lecture Notes in Mathematics 415, 78-84, Springer, Berlin.

Browne, P.J. (1977a). Abstract multiparameter theory I, J.Math.Anal.Appl.60, 259-273.

Browne, P.J. (1977b). Abstract multiparameter theory II, J.Math.Anal.Appl.60, 274-279.

Browne, P.J. (1977c). A completeness theorem for a nonlinear multiparameter eigenvalue problem, J.Differential Equations 23, 285-292.

Browne, P.J. (1978). The interlacing of eigenvalues for periodic multi-parameter problems, Proc.Roy.Soc.Edinburgh 80A, 357-362.

Browne, P.J. (1980). Abstract multiparameter theory III, J.Math.Anal.Appl.73, 561-567.

Browne, P.J. (1982). Multiparameter problems: the last decade, Proc. Conference on Ordinary and Partial Differential Equations Dundee 1982, Lecture Notes in Mathematics 904, 95-109, Springer, Berlin.

Browne, P.J. and Sleeman, B.D. (1978). Solvability of a linear operator system II, Quaestiones Mathematicae 3, 155-165.

Browne, P.J. and Sleeman, B.D. (1979a). Analytic perturbation of multiparameter eigenvalue problems, Quart.J.Math.Oxford (2) 30, 257-263.

Browne, P.J. and Sleeman, B.D. (1979b). Nonlinear multiparameter Sturm-Liouville problems, J.Differential Equations 34, 139-146.

Browne, P.J. and Sleeman, B.D. (1979c). Stability regions for multi-parameter systems of periodic second order ordinary differential equations, Proc.Roy.Soc.Edinburgh 84A, 249-257.

Browne, P.J. and Sleeman, B.D. (1979d). Regular multiparameter eigenvalue problems with several parameters in the boundary conditions, J.Math.Anal. Appl.72, 29-33.

Browne, P.J. and Sleeman, B.D. (1980a). Bifurcation from eigenvalues in non-linear multiparameter Sturm-Liouville problems, Glasgow Math.J.21, 85-90.

Browne, P.J. and Sleeman, B.D. (1980b). Applications of multiparameter spectral theory to special functions, Math.Proc.Camb.Phil.Soc.87, 275-283.

Browne, P.J. and Sleeman, B.D. (1980c). Non-linear multiparameter eigenvalue problems for ordinary differential equations, J.Math.Anal.Appl.77, 425-432.

Browne, P.J. and Sleeman, B.D. (1981). Non-linear multiparameter eigenvalue problems, Quaestiones Mathematicae 4, 277-283.

Browne, P.J. and Sleeman, B.D. (1982). A numerical technique for multiparameter eigenvalue problems, IMA J.Numerical Analysis 2, 451-457.

Browne, P.J. and Sleeman, B.D. (1983). Multiparameter spectral theory for symmetric relations, Quart.J.Math.Oxford (2) 34, 151-165.

Browne, P.J. and Sleeman, B.D. (1985). Inverse multi-parameter eigenvalue problems for matrices, Proc.Roy.Soc.Edinburgh 100A, 29-38.

Browne, P.J. and Sleeman, B.D. (1988). Multiparameter deficiency index theory, Applicable Anal.29,45-54.

Camp, C.C. (1928). An expansion involving P inseparable parameters associated with a partial differential equation, Amer.J.Math.50, 259-268.

Camp, C.C. (1930). On multiparameter expansions associated with a differential system and auxiliary conditions at several points in each variable, Amer.J. Math.60, 447-452.

Carmichael, R.D. (1921a). Boundary value and expansion problems: algebraic basis of the theory, Amer.J.Math.43, 69-101.

Carmichael, R.D. (1921b). Boundary value and expansion problems:formulation of various transcendental problems, Amer.J.Math.43, 232-270.

Carmichael, R.D. (1922). Boundary value and expansion problems:oscillatory, comparison and expansion problems, Amer.J.Math.44, 129-152.

Coddington, E.A. and Levinson, N. (1955). Theory of ordinary differential equations, McGraw Hill, New York.

Collatz, L. (1968). Multiparametric eigenvalue problems in inner-product spaces, J.Comp.System.Sci.2, 333-341.

Cordes, H.O. (1953). Separation der Variablen in Hilbertschen Räumen, Math. Ann.125, 401-434.

Cordes, H.O. (1954). Der Entwicklungssatz nach Produkten bei singulären Eigenwertproblemen partieller Differentialgleichungen, die durch Separation zerfallen, Nachr.Akad.Wiss.Göttingen, 51-69.

Cordes, H.O. (1954/55a). über die Spektralzerlegung von hypermaximalen Operatoren, die durch Separation der Variablen zerfallen I, Math.Ann.128, 257-289.

Cordes, H.O. (1954/55b). über die Spektralzerlegung von hypermaximalen Operatoren, die durch Separation der Variablen zerfallen II, Math.Ann.128, 373-411.

Deimling, K. (1985). Nonlinear functional analysis, Springer, Berlin-Heidelberg-New York-Tokyo.

Dixon, A.C. (1902). Expansions by means of Lamé's functions, Proc.London Math. Soc.(1) 35, 162-197.

Dixon, A.C. (1907). Harmonic expansions of functions of two variables, Proc.London Math.Soc.(2) 5, 411-478.

Doole, H.P. (1931). A certain multiparameter expansion, Bull.Amer.Math.Soc.37, 439-446.

Dunford, N. and Schwartz, J.T. (1963). Linear operators, Part II, Wiley Interscience, New York.

Erdélyi, A.;Magnus, W.;Oberhettinger, F. and Tricomi, F.G. (1955). Higher transcendental functions Vol. III, McGraw Hill, New York-Toronto-London.

Etgen, G.J. and Tefteller, S.C. (1979a). A two parameter boundary problem for a second order differential system, Ann.Mat.Pura.Appl.120 (Ser.4), 293-303.

Etgen, G.J. and Tefteller, S.C. (1979b). Three point boundary problems for second order differential equations.

Faierman, M. (1969). The completeness and expansion theorem associated with the multiparameter eigenvalue problem in ordinary differential equations, J.Differential Equations 5, 197-213.

Faierman, M. (1971). On a perturbation in a two-parameter ordinary differential equation of the second order, Canad.Math.Bull.14, 25-33.

Faierman, M. (1972a). An oscillation theorem for a one-parameter ordinary differential equation of the second order, J.Differential equations 11, 10-37.

Faierman, M. (1972b). Asymptotic formulae for the eigenvalues of a two-parameter ordinary differential equation of the second order, Trans.Amer. Math.Soc.168, 1-52.

Faierman, M. (1974). The expansion theorem in multi-parameter Sturm-Liouville theory, Lecture Notes in Mathematics 415, 137-142, Springer, Berlin.

Faierman, M. (1975a). Asymptotic formulae for the eigenvalues of a two-parameter system of ordinary differential equations of the second order, Canad.Math.Bull.17, 657-665.

Faierman, M. (1975b). A note on Klein's oscillation theorem for periodic boundary conditions, Canad.Math.Bull.17, 749-755.

Faierman, M. (1977). On the distribution of the eigenvalues of a two-parameter system of ordinary differential equations of the second order, SIAM J.Math. Anal.8, 854-870.

Faierman, M. (1978). Eigenfunction expansions associated with a two-parameter system of differential equations, Proc.Roy.Soc.Edinburgh 81A, 79-93.

Faierman, M. (1979a). Distribution of eigenvalues of a two-parameter system of differential equations, Trans.Amer.Math.Soc.247, 45-86.

Faierman, M. (1979b). An oscillation theorem for a two parameter system of differential equations, Quaestiones Mathematicae 3, 313-321.

Faierman, M. (1980). Bounds for the eigenfunctions of a two-parameter system of differential equations of the second order, Pacific J.Math. 90, 333-345.

Faierman, M. (1981a). An eigenfunction expansion associated with a two-parameter system of differential equations I, Proc.Roy.Soc.Edinburgh 89A, 143-155.

Faierman, M. (1981b). An eigenfunction expansion associated with a two-parameter system of differential equations, in Spectral Theory of Differential Operators, I.W.Knowles and R.T.Lewis (editors), North-Holland, 169-172.

Faierman, M. (1982a). An eigenfunction expansion associated with a two-parameter system of differential equations II, Proc.Roy.Soc.Edinburgh 92A, 87-93.

Faierman, M. (1982b). An oscillation theory for a two-parameter system of differential equations with periodic boundary conditions, Quaestiones Mathematicae 5, 107-118.

Faierman, M. (1983). An eigenfunction expansion associated with a two-parameter system of differential equations III, Proc.Roy.Soc.Edinburgh 93A, 189-195.

Faierman, M. (1985). The eigenvalues of a multiparameter system of differential equations, Applicable Anal.19, 275-290.

Faierman, M. (1986). Expansions in eigenfunctions of a two-parameter system of differential equations, Quaestiones Math.10, 135-152.

Faierman, M. (1987). Expansions in eigenfunctions of a two-parameter system of differential equations II, Quaestiones Math.10, 217-249.

Faierman, M. and Roach, G.F. (1987). Linear elliptic eigenvalue problems involving an indefinite weight, J.Math.Anal.Appl.126, 517-528.

Faierman, M. and Roach, G.F. (1988a). Full and partial-range eigenfunction expansions for a multiparameter system of differential equations, Applicable Anal.28, 15-37.

Faierman, M. and Roach, G.F. (1988b). Eigenfunction expansions associated with a multiparameter system of differential equations, submitted.

Gregus, N.; Neuman, F. and Arscott, F. (1971). Three point boundary value problems in differential equations, J.London Math.Soc.3, 429-436.

Guseinov, G.S. (1980). Eigenfunction expansion of multiparameter differential and difference equations with periodic coefficients, Soviet Math.Dokl.22, 201-205.

Hadeler, K.P. (1967). Mehrparametrige und nichtlineare Eigenwertaufgaben, Arch.Rational Mech.Anal.27, 306-328.

Hadeler, K.P. (1969). Einige Anwendungen mehrparametriger Eigenwertaufgaben, Numer.Math.13, 285-292.

Hargrave, B.A. and Sleeman, B.D. (1974a). The numerical solution of two-parameter eigenvalue problems with an application to the problem of diffraction by a plane angular sector, J.Inst.Maths.Applics.14, 9-22.

Hargrave, B.A. and Sleeman, B.D. (1974b). Uniform asymptotic expansions for ellipsoidal wave functions, J.Inst.Maths.Applics.14, 31-40.

Hargrave, B.A. and Sleeman, B.D. (1975). Asymptotic evaluation of certain integral formulae for ellipsoidal wave functions, Proc.Roy.Soc.Edinburgh 72A, 257-269.

Hausdorff, F. (1919). Der Wertvorrat einer Bilinearform, Math.Z.3, 314-316.

Hilb, E. (1907a). Eine Erweiterung des Kleinschen Oszillationstheorems, Jahresb.d.D.Math.-Ver.16, 279-285.

Hilb, E. (1907b). Die Reihenentwicklungen der Potentialtheorie, Math.Ann.63, 38-53.

Hilbert, D. (1912). Grundzüge einer allgemeinen Theorie der linearen Integral-gleichungen, Teubner, Leipzig.

Howe, A. (1971). Klein's oscillation theorem for periodic boundary conditions, Can.J.Math.23, 699-703.

Ince, E. (1926). Ordinary differential equations, Dover reprint, New York 1956.

Isaev, H.A. (1976). On multiparameter spectral theory, Soviet Math.Dokl.17, 1004-1007.

Isaev, H.A. (1980). On root elements of multiparameter spectral theory, Soviet Math.Dokl.21, 127-130..

Isaev, H.A. (1981a). Expansions in eigenfunctions of self-adjoint singular multiparameter differential operators, Soviet Math.Dokl.24, 326-330.

Isaev, H.A. (1981b). On the theory of deficiency indices for multiparameter differential operators of Sturm-Liouville type, Soviet Math.Dokl.24, 580-583.

Isaev, H.A. (1983). Genetic operators and multiparameter spectral theory, Soviet Math.Dokl.27, 149-152.

Isaev, H.A. (1985). Lectures on multiparameter theory, Department of Mathe-matics and Statistics, The University of Calgary.

Jörgens, K. and Rellich, F. (1976). Eigenwerttheorie gewöhnlicher Differential-gleichungen, Springer, Berlin-Heidelberg-New York.

Källström, A. and Sleeman, B.D. (1974). A multi-parameter Sturm Liouville problem, Proc.Conference on Ordinary and Partial Differential Equations, Lecture Notes in Mathematics 415, 394-401, Springer, Berlin.

Källström, A. and Sleeman, B.D. (1974/75a). An abstract relation for multi-parameter eigenvalue problems, Proc.Roy.Soc.Edinburgh 74A, 135-143.

Källström, A. and Sleeman, B.D. (1974/75b). A left definite multiparameter eigenvalue problem in ordinary differential equations, Proc.Roy.Soc. Edinburgh 74A, 145-155.

Källström, A. and Sleeman, B.D. (1975). An abstract multiparameter spectral theory, Univ.of Dundee Math.Report 75:2.

Källström, A. and Sleeman, B.D. (1976). Solvability of a linear operator system, J.Math.Anal.Appl.55, 785-793.

Källström, A. and Sleeman, B.D. (1977). Multiparameter spectral theory, Ark. Mat.15, 93-99.

Kalnins, E.G. (1986). Separation of variables for Riemannian spaces of constant curvature, Pitman.

Kato, T. (1966). Perturbation theory for linear operators, Springer, Berlin.

Klein, F. (1881). über Körper, welche von confocalen Flächen zweiten Grades begrenzt sind, Math.Ann.18, 410-427.

Mc Ghee, D.F. (1982). Multiparameter problems and joint spectra, Proc.Roy.Soc. Edinburgh 93A, 129-135.

Mc Ghee, D.F. and Picard, R.H. (1988). Cordes' two parameter spectral representation theory, Pitman.

Mc Ghee, D.F. and Roach, G.F. (1981). The spectrum of multiparameter problems in Hilbert space, Proc.Roy.Soc.Edinburgh 91A, 31-42.

Meixner, J. and Schäfke, F.W. (1954). Mathieusche Funktionen und Sphäroidfunktionen, Springer, Berlin.

Meixner, J.; Schäfke, F.W. and Wolf, G. (1980). Mathieu functions and spherical functions and their mathematical foundation, Lecture Notes in Mathematics 837, Springer, Berlin.

Miller, W. Jr. (1968). Lie theory and special functions, Academic Press.

Murray, F.J. and von Neumann, J. (1936). On rings of operators, Ann.Math.37, 116-229.

Pell, Anna J. (1922). Linear equations with two parameters, Trans.Amer.Math. Soc.23, 198-211.

Reid, W.T. (1971). Ordinary differential equations, Wiley, New York.

Richardson, R.G. (1912). Theorems of oscillation for two linear differential equations of the second order with two parameters, Trans.Amer.Math.Soc.13, 22-34.

Richardson, R.G. (1912/13). über die notwendigen und hinreichenden Bedingungen für das Bestehen eines Kleinschen Oszillations Theorems, Math.Ann.73, 289-304.

Riesz, F. and Nagy, B. (1972). Functional Analysis, Ungar, New York.

Roach, G.F. (1974). Transform theorems for multiparameter problems in Hilbert space, NAM-Bericht 14.

Roach, G.F. (1976). A Fredholm theory for multiparametric problems, Nieuw Arch. Wisk.24, 49-76.

Roach, G.F. (1977). Variational methods for multiparametric problems, ISNM 38, Birkhäuser Verlag.

Roach, G.F. (1982). Symmetry groups and multiparameter problems, Zeszyty Nauk. Politech.Lodz Math.14, 21-39.

Roach, G.F. (Editor) (1984). Multiparameter problems, Shiva Mathematics Series 8, Nantwich.

Roach, G.F. and Sleeman, B.D. (1976). Generalized multiparameter spectral theory, Proc.Conference on function theoretic methods in partial differential equations, Lecture Notes in Mathematics 561, 394-411, Springer, Berlin.

Roach, G.F. and Sleeman, B.D. (1977). On the spectral theory of operator bundles, Applicable Analysis 7, 1-14.

Roach, G.F. and Sleeman, B.D. (1978). Coupled systems of multiparameter eigenvalue problems, Proc.Roy.Soc.Edinburgh 80A, 223-234.

Roach G.F. and Sleeman, B.D. (1979). On the spectral theory of operator bundles II, Applicable Anal.9, 29-36.

Rynne G.F. and Sleeman, B.D. (1983). Bloch waves and multiparameter spectral theory, Proc.Roy.Soc.Edinburgh 95A, 73-93.

Sack, R.A. (1972). Variational solutions for eigenvalues of single and coupled Lame equations, J.Inst.Math.Applics.10, 279-288.

Sansone, G. (1959). Orthogonal functions, Interscience, New York.

Schäfke, R. and Volkmer, H. (1985). A note on the paper "Existence Conditions for eigenvalue problems generated by compact multiparameter operators" by P.Binding, P.J.Browne and L.Turyn, Proc.Roy.Soc.Edinburgh 101A, 147-148.

Schäfke, R. and Volkmer, H. (1988). Bounds for the eigenfunctions of multiparameter Sturm-Liouville systems, submitted.

Schmidt, D. and Wolf, G. (1979). A method of generating integral relations by the simultaneous separability of generalized Schrödinger equations, SIAM J.Math.Anal.10, 823-838.

Sleeman, B.D. (1971a). Multiparameter eigenvalue problems in ordinary differential equations, Bull.Inst.Politehn.Iasi 17, 51-60.

Sleeman, B.D. (1971b). The two-parameter Sturm-Liouville problem for ordinary differential equations, Proc.Roy.Soc.Edinburgh 69A, 139-148.

Sleeman, B.D. (1972a). Multiparameter eigenvalue problems and K-linear operators, Proc.Conference on Ordinary and Partial Differential Equations, Lecture Notes in Mathematics 280, 347-353, Springer, Berlin.

Sleeman, B.D. (1972b). The two-parameter Sturm-Liouville problem for ordinary differential equations II, Proc.Amer.Math.Soc.34, 165-170.

Sleeman, B.D. (1973a). Completeness and expansion theorems for a two-parameter eigenvalue problem in ordinary differential equations using variational principles, J.Lond.Math.Soc.6, 705-712.

Sleeman, B.D. (1973b). Singular linear differential operators with many parameters, Proc.Roy.Soc.Edinburgh 71A, 199-232.

Sleeman, B.D. (1974a). Some aspects of multi-parameter spectral theory, Proc. Conference in Spectral Theory and asymptotics of differential equations, Scheveningen, North Holland Mathematics Studies 13, 81-94.

Sleeman, B.D. (1974b). Left definite multiparameter eigenvalue problems, Proc.Symp.on spectral theory and differential equations, Lecture Notes in Mathematics 448, 307-321, Springer, Berlin.

Sleeman, B.D. (1978a). Multiparameter spectral theory in Hilbert space, Pitman, London.

Sleeman, B.D. (1978b). Multiparameter spectral theory in Hilbert space, J.Math. Anal.Appl.65, 511-530.

Sleeman, B.D. (1978c). Multiparameter periodic differential equations, Proc. Conf.on the Theory of Ordinary and Partial Differential Equations, Lecture Notes in Mathematics 827, 229-250, Springer, Berlin.

Sleeman, B.D. (1979). Klein oscillation theorems for multiparameter eigenvalue problems in ordinary differential equations, Nieuw Arch.Wisk.27, 341-362.

Stäckel, P. (1893a). Sur une classe de problèmes de dynamique, Comptes Rendus 116, 485-487.

Stäckel, P. (1893b). Sur des problèmes de dynamique, qui se reduisent a des quadratures, Comptes Rendus 116, 1284-1286.

Stäckel, P. (1893c). über die Bewegung eines Punktes in einer n-fachen Mannigfaltigkeit, Math.Ann 42, 537-563.

Stieltjes, T.J. (1885). Sur certains polynomes qui vérifient une équation différentielle linéaire du second ordre, et sur la théorie des fonctions de Lamé, Acta math.6, 312-326.

Turyn, L. (1980). Sturm-Liouville problems with several parameters, J.Differential Equations 38, 239-259.

Turyn, L. (1983). Perturbation of linked eigenvalue problems, J.Nonlinear Analysis 7, 35-40.

Volkmer, H. (1982). On multiparameter theory, J.Math.Anal.Appl.86, 44-53.

Volkmer, H. (1984a). On the completeness of eigenvectors of right definte multiparameter problems, Proc.Roy.Soc.Edinburgh 96A, 69-78.

Volkmer, H. (1984b). Eigenvector expansion in multiparameter eigenvalue problems, Shiva Mathematics Series 8, 93-101.

Volkmer, H. (1986). On the minimal eigenvalue of a positive definite operator determinant, Proc.Roy.Soc.Edinburgh 103A, 201-208.

Volkmer, H. (1987). On an expansion theorem of F.V.Atkinson and P.Binding, SIAM J.Math.Anal. 18, 1669-1680.

Volkmer, H. (1988a). The expansion of a holomorphic function in a Laplace series, submitted.

Volkmer, H. (1988b). The expansion of a holomorphic function in a series of Lamé products, preprint.

Weinstein, A. and Stenger, W. (1972). Methods of intermediate problems for eigenvalues, Academic Press, New York.

Whittaker, E.T. and Watson, G.N (1927). A course of modern analysis, Cambridge.

Yoshida, K. (1971). Functional Analysis, Springer, Berlin-Heidelberg-New York.

Yoshikawa, J. (1910). Ein zweiparametriges Oszillationstheorem, Nachr.Ges.Wiss. Göttingen, 586-594.

INDEX

LECTURE NOTES IN MATHEMATICS

Edited by A. Dold and B. Eckmann

Some general remarks on the publication of monographs and seminars

In what follows all references to monographs, are applicable also to multiauthorship volumes such as seminar notes.

§1. Lecture Notes aim to report new developments - quickly, informally, and at a high level. Monograph manuscripts should be reasonably self-contained and rounded off. Thus they may, and often will, present not only results of the author but also related work by other people. Furthermore, the manuscripts should provide sufficient motivation, examples and applications. This clearly distinguishes Lecture Notes manuscripts from journal articles which normally are very concise. Articles intended for a journal but too long to be accepted by most journals, usually do not have this "lecture notes" character. For similar reasons it is unusual for Ph.D. theses to be accepted for the Lecture Notes series.

Experience has shown that English language manuscripts achieve a much wider distribution.

§2. Manuscripts or plans for Lecture Notes volumes should be submitted either to one of the series editors or to Springer-Verlag, Heidelberg. These proposals are then refereed. A final decision concerning publication can only be made on the basis of the complete manuscripts, but a preliminary decision can usually be based on partial information: a fairly detailed outline describing the planned contents of each chapter, and an indication of the estimated length, a bibliography, and one or two sample chapters - or a first draft of the manuscript. The editors will try to make the preliminary decision as definite as they can on the basis of the available information.

3. Lecture Notes are printed by photo-offset from typed copy delivered in camera-ready form by the authors. Springer-Verlag provides technical instructions for the preparation of manuscripts, and will also, on request, supply special staionery on which the prescribed typing area is outlined. Careful preparation of the manuscripts will help keep production time short and ensure satisfactory appearance of the finished book. Running titles are not required; if however they are considered necessary, they should be uniform in appearance. We generally advise authors not to start having their final manuscripts specially tpyed beforehand. For professionally typed manuscripts, prepared on the special stationery according to our instructions, Springer-Verlag will, if necessary, contribute towards the typing costs at a fixed rate.

The actual production of a Lecture Notes volume takes 6-8 weeks.

.../...

§4. Final manuscripts should contain at least 100 pages of mathematical text and should include

- a table of contents
- an informative introduction, perhaps with some historical remarks. It should be accessible to a reader not particularly familiar with the topic treated.
- a subject index; this is almost always genuinely helpful for the reader.

§5. Authors receive a total of 50 free copies of their volume, but no royalties. They are entitled to purchase further copies of their book for their personal use at a discount of 33.3 %, other Springer mathematics books at a discount of 20 % directly from Springer-Verlag.

Commitment to publish is made by letter of intent rather than by signing a formal contract. Springer-Verlag secures the copyright for each volume.

Vol. 1201: Curvature and Topology of Riemannian Manifolds. Proceedings, 1985. Edited by K. Shiohama, T. Sakai and T. Sunada. VII, 336 pages. 1986.

Vol. 1202: A. Dür, Möbius Functions, Incidence Algebras and Power Series Representations. XI, 134 pages. 1986.

Vol. 1203: Stochastic Processes and Their Applications. Proceedings, 1985. Edited by K. Itô and T. Hida. VI, 222 pages. 1986.

Vol. 1204: Séminaire de Probabilités XX, 1984/85. Proceedings. Edité par J. Azéma et M. Yor. V, 639 pages. 1986.

Vol. 1205: B.Z. Moroz, Analytic Arithmetic in Algebraic Number Fields. VII, 177 pages. 1986.

Vol. 1206: Probability and Analysis, Varenna (Como) 1985. Seminar. Edited by G. Letta and M. Pratelli. VIII, 280 pages. 1986.

Vol. 1207: P.H. Bérard, Spectral Geometry: Direct and Inverse Problems. With an Appendix by G. Besson. XIII, 272 pages. 1986.

Vol. 1208: S. Kaijser, J.W. Pelletier, Interpolation Functors and Duality. IV, 167 pages. 1986.

Vol. 1209: Differential Geometry, Peñíscola 1985. Proceedings. Edited by A.M. Naveira, A. Ferrández and F. Mascaró. VIII, 306 pages. 1986.

Vol. 1210: Probability Measures on Groups VIII. Proceedings, 1985. Edited by H. Heyer. X, 386 pages. 1986.

Vol. 1211: M.B. Sevryuk, Reversible Systems. V, 319 pages. 1986.

Vol. 1212: Stochastic Spatial Processes. Proceedings, 1984. Edited by P. Tautu. VIII, 311 pages. 1986.

Vol. 1213: L.G. Lewis, Jr., J.P. May, M. Steinberger, Equivariant Stable Homotopy Theory. IX, 538 pages. 1986.

Vol. 1214: Global Analysis – Studies and Applications II. Edited by Yu.G. Borisovich and Yu.E. Gliklikh. V, 275 pages. 1986.

Vol. 1215: Lectures in Probability and Statistics. Edited by G. del Pino and R. Rebolledo. V, 491 pages. 1986.

Vol. 1216: J. Kogan, Bifurcation of Extremals in Optimal Control. VIII, 106 pages. 1986.

Vol. 1217: Transformation Groups. Proceedings, 1985. Edited by S. Jackowski and K. Pawalowski. X, 396 pages. 1986.

Vol. 1218: Schrödinger Operators, Aarhus 1985. Seminar. Edited by E. Balslev. V, 222 pages. 1986.

Vol. 1219: R. Weissauer, Stabile Modulformen und Eisensteinreihen. III, 147 Seiten. 1986.

Vol. 1220: Séminaire d'Algèbre Paul Dubreil et Marie-Paule Malliavin. Proceedings, 1985. Edité par M. P. Malliavin. IV, 200 pages. 1986.

Vol. 1221: Probability and Banach Spaces. Proceedings, 1985. Edited by J. Bastero and M. San Miguel. XI, 222 pages. 1986.

Vol. 1222: A. Katok, J.-M. Strelcyn, with the collaboration of F. Ledrappier and F. Przytycki, Invariant Manifolds, Entropy and Billiards; Smooth Maps with Singularities. VIII, 283 pages. 1986.

Vol. 1223: Differential Equations in Banach Spaces. Proceedings, 1985. Edited by A. Favini and E. Obrecht. VIII, 299 pages. 1986.

Vol. 1224: Nonlinear Diffusion Problems, Montecatini Terme 1985. Seminar. Edited by A. Fasano and M. Primicerio. VIII, 188 pages. 1986.

Vol. 1225: Inverse Problems, Montecatini Terme 1986. Seminar. Edited by G. Talenti. VIII, 204 pages. 1986.

Vol. 1226: A. Buium, Differential Function Fields and Moduli of Algebraic Varieties. IX, 146 pages. 1986.

Vol. 1227: H. Helson, The Spectral Theorem. VI, 104 pages. 1986.

Vol. 1228: Multigrid Methods II. Proceedings, 1985. Edited by W. Hackbusch and U. Trottenberg. VI, 336 pages. 1986.

Vol. 1229: O. Bratteli, Derivations, Dissipations and Group Actions on C*-algebras. IV, 277 pages. 1986.

Vol. 1230: Numerical Analysis. Proceedings, 1984. Edited by J.-P. Hennart. X, 234 pages. 1986.

Vol. 1231: E.-U. Gekeler, Drinfeld Modular Curves. XIV, 107 pages. 1986.

Vol. 1232: P.C. Schuur, Asymptotic Analysis of Soliton Problems. VIII, 180 pages. 1986.

Vol. 1233: Stability Problems for Stochastic Models. Proceedings, 1985. Edited by V.V. Kalashnikov, B. Penkov and V.M. Zolotarev. VI, 223 pages. 1986.

Vol. 1234: Combinatoire énumérative. Proceedings, 1985. Edité par G. Labelle et P. Leroux. XIV, 387 pages. 1986.

Vol. 1235: Séminaire de Théorie du Potentiel, Paris, No. 8. Directeurs: M. Brelot, G. Choquet et J. Deny. Rédacteurs: F. Hirsch et G. Mokobodzki. III, 209 pages. 1987.

Vol. 1236: Stochastic Partial Differential Equations and Applications. Proceedings, 1985. Edited by G. Da Prato and L. Tubaro. V, 257 pages. 1987.

Vol. 1237: Rational Approximation and its Applications in Mathematics and Physics. Proceedings, 1985. Edited by J. Gilewicz, M. Pindor and W. Siemaszko. XII, 350 pages. 1987.

Vol. 1238: M. Holz, K.-P. Podewski and K. Steffens, Injective Choice Functions. VI, 183 pages. 1987.

Vol. 1239: P. Vojta, Diophantine Approximations and Value Distribution Theory. X, 132 pages. 1987.

Vol. 1240: Number Theory, New York 1984–85. Seminar. Edited by D.V. Chudnovsky, G.V. Chudnovsky, H. Cohn and M.B. Nathanson. V, 324 pages. 1987.

Vol. 1241: L. Gårding, Singularities in Linear Wave Propagation. III, 125 pages. 1987.

Vol. 1242: Functional Analysis II, with Contributions by J. Hoffmann-Jørgensen et al. Edited by S. Kurepa, H. Kraljević and D. Butković. VII, 432 pages. 1987.

Vol. 1243: Non Commutative Harmonic Analysis and Lie Groups. Proceedings, 1985. Edited by J. Carmona, P. Delorme and M. Vergne. V, 309 pages. 1987.

Vol. 1244: W. Müller, Manifolds with Cusps of Rank One. XI, 158 pages. 1987.

Vol. 1245: S. Rallis, L-Functions and the Oscillator Representation. XVI, 239 pages. 1987.

Vol. 1246: Hodge Theory. Proceedings, 1985. Edited by E. Cattani, F. Guillén, A. Kaplan and F. Puerta. VII, 175 pages. 1987.

Vol. 1247: Séminaire de Probabilités XXI. Proceedings. Edité par J. Azéma, P.A. Meyer et M. Yor. IV, 579 pages. 1987.

Vol. 1248: Nonlinear Semigroups, Partial Differential Equations and Attractors. Proceedings, 1985. Edited by T.L. Gill and W.W. Zachary. IX, 185 pages. 1987.

Vol. 1249: I. van den Berg, Nonstandard Asymptotic Analysis. IX, 187 pages. 1987.

Vol. 1250: Stochastic Processes – Mathematics and Physics II. Proceedings 1985. Edited by S. Albeverio, Ph. Blanchard and L. Streit. VI, 359 pages. 1987.

Vol. 1251: Differential Geometric Methods in Mathematical Physics. Proceedings, 1985. Edited by P.L. García and A. Pérez-Rendón. VII, 300 pages. 1987.

Vol. 1252: T. Kaise, Représentations de Weil et GL_2 Algèbres de division et GL_n. VII, 203 pages. 1987.

Vol. 1253: J. Fischer, An Approach to the Selberg Trace Formula via the Selberg Zeta-Function. III, 184 pages. 1987.

Vol. 1254: S. Gelbart, I. Piatetski-Shapiro, S. Rallis. Explicit Constructions of Automorphic L-Functions. VI, 152 pages. 1987.

Vol. 1255: Differential Geometry and Differential Equations. Proceedings, 1985. Edited by C. Gu, M. Berger and R.L. Bryant. XII, 243 pages. 1987.

Vol. 1256: Pseudo-Differential Operators. Proceedings, 1986. Edited by H.O. Cordes, B. Gramsch and H. Widom. X, 479 pages. 1987.

Vol. 1257: X. Wang, On the C*-Algebras of Foliations in the Plane. V, 165 pages. 1987.

Vol. 1258: J. Weidmann, Spectral Theory of Ordinary Differential Operators. VI, 303 pages. 1987.

Vol. 1259: F. Cano Torres, Desingularization Strategies for Three-Dimensional Vector Fields. IX, 189 pages. 1987.

Vol. 1260: N.H. Pavel, Nonlinear Evolution Operators and Semi-groups. VI, 285 pages. 1987.

Vol. 1261: H. Abels, Finite Presentability of S-Arithmetic Groups. Compact Presentability of Solvable Groups. VI, 178 pages. 1987.

Vol. 1262: E. Hlawka (Hrsg.), Zahlentheoretische Analysis II. Seminar, 1984–86. V, 158 Seiten. 1987.

Vol. 1263: V.L. Hansen (Ed.), Differential Geometry. Proceedings, 1985. XI, 288 pages. 1987.

Vol. 1264: Wu Wen-tsün, Rational Homotopy Type. VIII, 219 pages. 1987.

Vol. 1265: W. Van Assche, Asymptotics for Orthogonal Polynomials. VI, 201 pages. 1987.

Vol. 1266: F. Ghione, C. Peskine, E. Sernesi (Eds.), Space Curves. Proceedings, 1985. VI, 272 pages. 1987.

Vol. 1267: J. Lindenstrauss, V.D. Milman (Eds.), Geometrical Aspects of Functional Analysis. Seminar. VII, 212 pages. 1987.

Vol. 1268: S.G. Krantz (Ed.), Complex Analysis. Seminar, 1986. VII, 195 pages. 1987.

Vol. 1269: M. Shiota, Nash Manifolds. VI, 223 pages. 1987.

Vol. 1270: C. Carasso, P.-A. Raviart, D. Serre (Eds.), Nonlinear Hyperbolic Problems. Proceedings, 1986. XV, 341 pages. 1987.

Vol. 1271: A.M. Cohen, W.H. Hesselink, W.L.J. van der Kallen, J.R. Strooker (Eds.), Algebraic Groups Utrecht 1986. Proceedings. XII, 284 pages. 1987.

Vol. 1272: M.S. Livšic, L.L. Waksman, Commuting Nonselfadjoint Operators in Hilbert Space. III, 115 pages. 1987.

Vol. 1273: G.-M. Greuel, G. Trautmann (Eds.), Singularities, Representation of Algebras, and Vector Bundles. Proceedings, 1985. XIV, 383 pages. 1987.

Vol. 1274: N. C. Phillips, Equivariant K-Theory and Freeness of Group Actions on C*-Algebras. VIII, 371 pages. 1987.

Vol. 1275: C.A. Berenstein (Ed.), Complex Analysis I. Proceedings, 1985–86. XV, 331 pages. 1987.

Vol. 1276: C.A. Berenstein (Ed.), Complex Analysis II. Proceedings, 1985–86. IX, 320 pages. 1987.

Vol. 1277: C.A. Berenstein (Ed.), Complex Analysis III. Proceedings, 1985–86. X, 350 pages. 1987.

Vol. 1278: S.S. Koh (Ed.), Invariant Theory. Proceedings, 1985. V, 102 pages. 1987.

Vol. 1279: D. Ieşan, Saint-Venant's Problem. VIII, 162 Seiten. 1987.

Vol. 1280: E. Neher, Jordan Triple Systems by the Grid Approach. XII, 193 pages. 1987.

Vol. 1281: O.H. Kegel, F. Menegazzo, G. Zacher (Eds.), Group Theory. Proceedings, 1986. VII, 179 pages. 1987.

Vol. 1282: D.E. Handelman, Positive Polynomials, Convex Integral Polytopes, and a Random Walk Problem. XI, 136 pages. 1987.

Vol. 1283: S. Mardešić, J. Segal (Eds.), Geometric Topology and Shape Theory. Proceedings, 1986. V, 261 pages. 1987.

Vol. 1284: B.H. Matzat, Konstruktive Galoistheorie. X, 286 pages. 1987.

Vol. 1285: I.W. Knowles, Y. Saitō (Eds.), Differential Equations and Mathematical Physics. Proceedings, 1986. XVI, 499 pages. 1987.

Vol. 1286: H.R. Miller, D.C. Ravenel (Eds.), Algebraic Topology. Proceedings, 1986. VII, 341 pages. 1987.

Vol. 1287: E.B. Saff (Ed.), Approximation Theory, Tampa. Proceedings, 1985–1986. V, 228 pages. 1987.

Vol. 1288: Yu. L. Rodin, Generalized Analytic Functions on Riemann Surfaces. V, 128 pages, 1987.

Vol. 1289: Yu. I. Manin (Ed.), K-Theory, Arithmetic and Geometry. Seminar, 1984–1986. V, 399 pages. 1987.

Vol. 1290: G. Wüstholz (Ed.), Diophantine Approximation and Transcendence Theory. Seminar, 1985. V, 243 pages. 1987.

Vol. 1291: C. Mœglin, M.-F. Vignéras, J.-L. Waldspurger, Correspondances de Howe sur un Corps p-adique. VII, 163 pages. 1987.

Vol. 1292: J.T. Baldwin (Ed.), Classification Theory. Proceedings, 1985. VI, 500 pages. 1987.

Vol. 1293: W. Ebeling, The Monodromy Groups of Isolated Singularities of Complete Intersections. XIV, 153 pages. 1987.

Vol. 1294: M. Queffélec, Substitution Dynamical Systems – Spectral Analysis. XIII, 240 pages. 1987.

Vol. 1295: P. Lelong, P. Dolbeault, H. Skoda (Réd.), Séminaire d'Analyse P. Lelong – P. Dolbeault – H. Skoda. Seminar, 1985/1986. VII, 283 pages. 1987.

Vol. 1296: M.-P. Malliavin (Ed.), Séminaire d'Algèbre Paul Dubreil et Marie-Paule Malliavin. Proceedings, 1986. IV, 324 pages. 1987.

Vol. 1297: Zhu Y.-I., Guo B.-y. (Eds.), Numerical Methods for Partial Differential Equations. Proceedings. XI, 244 pages. 1987.

Vol. 1298: J. Aguadé, R. Kane (Eds.), Algebraic Topology, Barcelona 1986. Proceedings. X, 255 pages. 1987.

Vol. 1299: S. Watanabe, Yu.V. Prokhorov (Eds.), Probability Theory and Mathematical Statistics. Proceedings, 1986. VIII, 589 pages. 1988.

Vol. 1300: G.B. Seligman, Constructions of Lie Algebras and their Modules. VI, 190 pages. 1988.

Vol. 1301: N. Schappacher, Periods of Hecke Characters. XV, 160 pages. 1988.

Vol. 1302: M. Cwikel, J. Peetre, Y. Sagher, H. Wallin (Eds.), Function Spaces and Applications. Proceedings, 1986. VI, 445 pages. 1988.

Vol. 1303: L. Accardi, W. von Waldenfels (Eds.), Quantum Probability and Applications III. Proceedings, 1987. VI, 373 pages. 1988.

Vol. 1304: F.Q. Gouvêa, Arithmetic of p-adic Modular Forms. VIII, 121 pages. 1988.

Vol. 1305: D.S. Lubinsky, E.B. Saff, Strong Asymptotics for Extremal Polynomials Associated with Weights on ℝ. VII, 153 pages. 1988.

Vol. 1306: S.S. Chern (Ed.), Partial Differential Equations. Proceedings. 1986. VI, 294 pages. 1988.

Vol. 1307: T. Murai, A Real Variable Method for the Cauchy Transform, and Analytic Capacity. VIII, 133 pages. 1988.

Vol. 1308: P. Imkeller, Two-Parameter Martingales and Their Quadratic Variation. IV, 177 pages. 1988.

Vol. 1309: B. Fiedler, Global Bifurcation of Periodic Solutions with Symmetry. VIII, 144 pages. 1988.

Vol. 1310: O.A. Laudal, G. Pfister, Local Moduli and Singularities. V, 117 pages. 1988.

Vol. 1311: A. Holme, R. Speiser (Eds.), Algebraic Geometry, Sundance 1986. Proceedings. VI, 320 pages. 1988.

Vol. 1312: N.A. Shirokov, Analytic Functions Smooth up to the Boundary. III, 213 pages. 1988.

Vol. 1313: F. Colonius, Optimal Periodic Control. VI, 177 pages 1988.

Vol. 1314: A. Futaki, Kähler-Einstein Metrics and Integral Invariants. IV, 140 pages. 1988.

Vol. 1315: R.A. McCoy, I. Ntantu, Topological Properties of Spaces o Continuous Functions. IV, 124 pages. 1988

Vol. 1316: H. Korezlioglu, A.S. Ustunel (Eds.), Stochastic Analysis an Related Topics. Proceedings, 1986. V, 371 pages. 1988.

Vol. 1317: J. Lindenstrauss, V.D. Milman (Eds.), Geometric Aspects Functional Analysis. Seminar, 1986–87. VII, 289 pages. 1988.

Vol. 1318: Y. Felix (Ed.). Algebraic Topology – Rational Homoto Proceedings. 1986. VIII, 245 pages. 1988

Vol. 1319: M. Vuorinen, Conformal Geometry and Quasiregu Mappings. XIX, 209 pages. 1988.